A Natural Curiosity

The Story of the Bell Museum

LANSING SHEPARD, DON LUCE,
BARBARA COFFIN, AND GWEN SCHAGRIN
WITH SALLY BRUMMEL, GEORGE WEIBLEN,
AND TIM WHITFELD

UNIVERSITY OF MINNESOTA PRESS
MINNEAPOLIS · LONDON

This publication was made possible in part by the people of Minnesota through a grant funded by an appropriation to the Minnesota Historical Society from the Minnesota Arts and Cultural Heritage Fund. Any views, findings, opinions, conclusions, or recommendations expressed in this publication are those of the authors and do not necessarily represent those of the State of Minnesota, the Minnesota Historical Society, or the Minnesota Historic Resources Advisory Committee.

The University of Minnesota Press gratefully acknowledges the generous assistance provided for the publication of this book from the Bell Museum.

All photographs appear courtesy of the Bell Museum unless credited otherwise.

Copyright 2022 by the Regents of the University of Minnesota

All rights reserved. No part of this publication may be reproduced, stored in a retrieval system, or transmitted, in any form or by any means, electronic, mechanical, photocopying, recording, or otherwise, without the prior written permission of the publisher.

Published by the University of Minnesota Press
111 Third Avenue South, Suite 290
Minneapolis, MN 55401-2520
http://www.upress.umn.edu

ISBN 978-1-5179-1036-5
A Cataloging-in-Publication record for this book is available from the Library of Congress.

Printed in Canada on acid-free paper

The University of Minnesota is an equal-opportunity educator and employer.

28 27 26 25 24 23 22   10 9 8 7 6 5 4 3 2 1

Contents

A Note to Readers VIII

Foreword IX
Ford W. Bell

Introduction 1

Bell Museum Timeline 6

1. A Museum Is Born, 1872–1940

Documenting Minnesota: The Geological and Natural History Survey 14

The Menage Expedition: How Orangutan Bones Landed in the Bell Museum Collections 19

Josephine Tilden: Paving the Way for Women in Science 21

Thomas Sadler Roberts: Naturalist, Doctor, Director 25

Making a Museum for the Public: The Early Dioramas 33

2. Growing an Institution, 1920s–1960s

The Many Talents of Walter Breckenridge 42

Early Public Education: Reaching "the whole people . . ." 47

James Ford Bell: The Man behind the Name 52

Heyday of the Dioramas: Windows into Nature 58

Taking Flight: The Artistic Journey of Francis Lee Jaques 66

3. Wildlife Explorations, 1940s–1980s

At the Poles: Arctic and Antarctic Research 76

The Bride Wore . . . Boots? 76

Migrations: The Life and Times of Dwain Warner 78

Tracking Nature: The Rise of Wildlife Telemetry 82

Mystery of the Missing Toads 86

4. THE MUSEUM IN THE ENVIRONMENTAL ERA, 1960s–1990s

Touch and See: Innovating Hands-On Learning 90
Public Programs: From Education to Engagement 94
Interpreting Nature: The Student Guide Program 99
From Student Guide to College Professor 104
Making Movies: Reaching a Bigger Audience 106
Honeybees on the Roof: Sweetening Science Education 110
Widening the Inquiry: Bringing Together Ecology, Evolution, and Behavior 114
Nature versus Nurture: Frank McKinney and the Evolution of Animal Behavior 120
Minnesota's Rarest: Naming the State's Endangered and Threatened Flora and Fauna 125
Flight of the Peregrine: Bud Tordoff and the Return of an Endangered Species 129
Art and Natural History: The Evolution of a Legacy 135
Science through the Lens of Art: Resident Artists at the Bell 140
Change Comes to the "Eternal" Museum: Temporary and Traveling Exhibits 144

5. REDISCOVERING THE COLLECTIONS, 1980s–2022

Collections Offer Clues to Environmental Challenges 150
A Botanical Treasure: The University of Minnesota Herbarium 155
The DNA Revolution Comes to the Bell Museum 160
Rethinking the Tree of Life 164
Bell Museum Scientists on the Global Stage 169
Biodiversity Research: Understanding Life's Threatened Diversity 172
One Hundred Years Later: Minnesota Updates Its Natural History Survey 178
Collections Go Online 181

6. A MUSEUM FOR THE TWENTY-FIRST CENTURY, 1990s–2022

Saving an Endangered Museum: Surviving and Thriving in a University Setting 186
From the Earth to the Cosmos: The Journey of Minnesota's Planetarium 190
The Road to a Reimagined Museum 196
Designing with Nature: The Bell Museum's New Home 203
Moving Minnesota: Dioramas in a New Habitat 209
The Experience: A Journey through Time 217

Afterword 229
Denise Young

Acknowledgments 230

APPENDIXES

    The Bell Museum Dioramas 232
    Selected Exhibitions at the Bell Museum 240
    Selected Publications of the Bell Museum 242

Selected References 245

Contributors 260

Index 263

A Note to Readers

Histories can be hard reads, particularly for those whose ancestors have been ignored or subjected to abuse and marginalization. The history of the Bell Museum is no exception. It is a product of a society that was, and still is, wracked by deep social inequities. The freedom, education, and economic opportunities of many groups—women, people of color, and Indigenous people among them—have been systematically restricted. The result? The museum's history is largely dominated by the activities of white males, a story too often repeated in institutions of science and conservation.

The issue of race, in particular, looms large in America. This country's mistreatment of its nonwhite citizens is a historic feature woven into the fabric of the nation from its founding. Indeed, the ground on which the University of Minnesota and the Bell Museum stand, like the rest of the state, was appropriated through treaties negotiated in bad faith with Indigenous peoples.

The Morrill Act of 1862 granted federal land to universities throughout the country to fund practical education in agriculture, science, and technology. The act is seen as a major force for helping to make American universities more democratic, but the source of this federal land is often overlooked. Much of the acreage granted to the University came from territory surrendered by the Dakota in the 1851 treaties at Mendota and Traverse des Sioux. This is just one example of how the influx of white immigrants benefited from the dispossession of Native people.

The collections of the Bell Museum, as established by the Minnesota Geological and Natural History Survey, remind us of how natural resources were cataloged to legitimize and enforce a foreign system of ownership in the name of science. While this book celebrates what we have learned from the study of preserved specimens, we must bear in mind how misappropriation and trauma are bound up with the enterprise of science as has been practiced in natural history museums.

All of this informs the Bell Museum as it looks to the future. The museum aims to serve society in addressing the global challenges we all face—challenges that are strongly tied to the environment. Solutions to climate change, population growth, pollution, biodiversity loss, and pandemic disease all require an understanding of how ecosystems work. But solutions also require recognition and understanding of how societal inequities have distorted the impacts of these problems. Whether from asthma from air pollution, toxic contamination from industries, or flooding from climate change, communities of color and people living in poverty suffer disproportionately. A more sustainable future can be achieved only through greater justice and equality.

Times are changing and so is the Bell Museum. Going forward, it is our great task to listen and learn and to teach about and help build an appreciation for the natural world in the ongoing struggle to repair and preserve what is, after all, the only home we have.

Foreword

Ford W. Bell

The history of the Bell Museum is a history of far-sighted individuals with big ideas who had important dreams for the future of their state.

In 1872, just fourteen years after Minnesota achieved statehood, the state legislature mandated a geological and natural history survey of the state, including the creation of a natural history museum. At that time, the country was still emerging from the trauma of the Civil War, and Minnesota legislators undoubtedly had a lot on their minds. But the country was embracing big ideas: the Metropolitan Museum of Art and Yellowstone National Park were also created in 1872. Minnesota was looking to the future.

In 1915, the University of Minnesota named Thomas Sadler Roberts, a physician, to be professor of ornithology and associate curator of the Zoological Museum. That appointment encouraged a partnership with one of his patients, James Ford Bell, my grandfather and a lover of the natural world. Dr. Roberts and my grandfather came from very different backgrounds—one a physician, the other a miller—but their reverence for the natural world, rooted in their shared Philadelphia heritage, influenced by the Quaker concern for the living world and a sense of responsibility for its care, brought them together.

In April 1938, the two friends, sitting in my grandfather's office, planned their most significant undertaking, one that had been mostly a dream for the previous thirty-three years: the first building dedicated solely for Minnesota's natural history museum. Today we celebrate another new Bell Museum building, including a state-of-the-art planetarium, an institution that will allow generations of Minnesotans to learn about, appreciate, and preserve the natural world that is a cherished legacy of our state.

Natural history museums play a unique role in our society today as we seek to understand and protect the precious biodiversity that supports life on Earth. They bring us face to face with the challenges of a warming planet, of resource scarcity, of emerging infectious diseases, and of growing threats to our air, water, soil, and habitats. Natural history museums promote knowledge, stimulate discussion, encourage conservation, and offer visitors a glimpse of the priceless natural world that surrounds and nurtures us all. I know that is what my grandfather would have wanted for the museum that bears his name.

Collections are the heart of museums. Photograph by Joe Szurszewski.

Introduction

A child picks up a maple leaf all crimson and gold, or a pine cone off a forest floor, or she spots a dragonfly glistening with dew on a cool morning in a meadow.

These encounters with objects from nature can open a window to wonder. She asks, "What is it? What does it do? Where did it come from?" She is told its name but also that there are others like it but different. Can she find those others? Here is the seed of our natural curiosity. The ancient Greeks would say that the Muses had been summoned—those goddesses of wonder and protectors of knowledge who bring inspiration to poets, artists, and even naturalists.

The word museum means the "place of the Muses," the place where we collect, name, document, classify, and protect objects of wonder. For much of human evolution, the need to identify nature around us was essential for survival. How do you tell the harmless snake from the one that could kill you? Which plant will treat your illness, and which will make you sick? Native peoples nearly everywhere have had this knowledge and have used it to thrive. But most of humanity goes through life knowing very little about the natural world, a world that we all ultimately depend on.

Some people see this as a problem. As the gifted Minnesota nature writer Paul Gruchow so eloquently wrote, "It is perhaps the quintessentially human character that we cannot know or love what we have not named." Natural history museums are the keepers of the names for the world's amazing natural diversity. They protect a record of that diversity and strive to understand its origins and functions. Their collections, if skillfully cared for and displayed, can be powerful objects of wonder that open the door to curiosity, exploration, and appreciation. The Bell Museum was born from this hope.

The 1872 legislation that founded the Minnesota Geological and Natural History Survey and the museum was not written by a member of the legislature but by William Folwell, the first president of the University of Minnesota. Trained as a mathematician, Folwell was a true believer in the power of education and the advancement of knowledge to drive progress and improve society. When he took on the presidency of the school in 1869, it had only eight faculty members and one hundred students, all housed in a single building—a Victorian stone hulk known as "Old Main."

Folwell was a man with big plans. He was determined to make the University a top-tier institution of learning and set out to do so by expanding its undergraduate offerings to include postgraduate education and professional programs as well as cultural facilities such as museums. When the Minnesota Legislature came to him and asked him to write the enabling legislation for a comprehensive survey of the state's natural resources to be run out of the University, Folwell jumped at the offer. Here was a way to further his vision. The legislation he wrote was rich in detail about what should be investigated, collected, and preserved and was specific about what was to happen to the collections. Collected specimens, he wrote, were "to be preserved for public inspection . . . , in the University of Minnesota, in rooms convenient of access and properly warmed, lighted, ventilated and furnished, and in charge of a proper scientific curator." Further, "in the employment of assistants in the said surveys the said board of regents shall at all times give the preference to the students and graduates of the University of Minnesota."

If Folwell was the museum's architect who supplied the blueprint, Newton Horace Winchell, the person he hired to turn his words into a reality, was the man who laid the museum's foundation. Winchell was a geologist and respected veteran of

Michigan's own state survey. He turned out to be equally ambitious in his plans for Minnesota's survey and museum. Given no money to fund the initiative, he orchestrated legislation to provide the survey and museum with ongoing revenue through the early years. He assembled what scientific expertise the University had, enlisting students and volunteers in the effort. He organized the survey's lines of inquiry—zoology, botany, and geology—and the collections into which the gathered specimens were placed. And he commandeered a room on the third floor of Old Main to house the collected birds, mammals, plants, amphibians, fish, insects, and rocks that began pouring in. Inauspicious as this small space was, Winchell saw it as the beginning of something grand, both a prized University institution and a valued center of learning for students and for the public.

The founding of the museum within a land-grant university has had a profound effect on its character and its work. As a public, land-grant institution, the University has a three-part mission: to conduct research and advance new knowledge, to teach students at the undergraduate and graduate levels, and to conduct outreach, spreading that knowledge and its benefits to the citizens of the state and the larger world. Pursuit of these missions at the Bell Museum has created a unique environment where researchers exploring the workings of natural systems have collaborated with artists creating new ways to express their affinity for nature, while educators exploring new ways of learning have used those collaborations to develop creative ways to impart that synergism to visitors.

This composite mission has been reflected in the people entrusted to carry it out. In the 1960s, Dwain Warner, the museum's first professional ornithologist, was experimenting with the use of radio-tracking technology for wildlife research, while the renowned artist Francis Lee Jaques was bringing the art of dioramas to full expression in his painted scenes of Minnesota habitats, and the educator Richard Barthelemy was trying out novel approaches to hands-on learning. Sometimes these missions would be expressed all in one person—such was the museum's fourth director, Walter Breckenridge. Breckenridge started his career as a taxidermist, earned a doctorate with his original research on Minnesota's amphibians and reptiles, and became a great champion for public outreach through his films, lectures, exhibitions, state park naturalist program, and his own artwork.

Keeping true to President Folwell's vision, University students have been engaged in all aspects of the museum, from conducting research in sometimes far-flung locales, to preparing

William Folwell served as president, faculty member, and head librarian of the University of Minnesota. University of Minnesota Archives.

Old Main, the first home of the University and its natural history museum. University of Minnesota Archives.

specimens for the collections, to teaching evening classes for the public, guiding school groups through the museum, helping to develop new exhibits and education programs, working at the front desk, and even making wax leaves for the dioramas!

Over the 150 years of the museum's existence, events great and small have swept the nation and world, shaping and reshaping science, art, and society. Like most museums, the Bell Museum has not been immune to these transitions. The museum has responded and adapted to changing trends and breakthroughs in science, technology, and informal education. Indeed, the museum has played a role in helping shape those larger forces.

While this book is a history of the Bell Museum, it is not a straight chronology of museum events and activities. Instead, it is organized into a series of stories focusing on particular aspects of the museum. Some stories feature a person who made a significant contribution to the museum's development, others feature programs that changed the direction of the museum, and still others highlight major scientific advancements in which the museum participated.

One of the key personalities featured in the book is former museum director Harrison "Bud" Tordoff. His leadership led to a major expansion and transformation of the museum's research and teaching, as well as to one of the most successful conservation programs—the restoration of the endangered peregrine falcon. Another story traces the museum's use of films, movies, and other media. From a multipart public television series to electronic field trips, the museum has used media to greatly expand its public outreach. Several stories reflect how the rise of the environmental movement in the 1960s and 1970s led to a flowering of activity in the museum that transformed its research, teaching, and public programs.

During the past 150 years, science has undergone major transformations. Biology in the nineteenth century was largely a descriptive process of documenting and classifying the diversity of organisms. As more came to be known about the natural world, scientific inquiry focused on understanding the functioning of those organisms, then on the functioning of communities of organisms. This ecological perspective was then challenged by the revolution of molecular biology, which threatened to relegate natural history museums to the historical waste bin. The story of how the Bell Museum dramatically reoriented the nature of its research to embrace this new technology, and then to use it to turn its fortunes around is largely the story of how natural history museums restored their role as important players in the advancement of science during the twenty-first century.

Natural history museums have long been associated with the rise of the American conservation movement. This movement started in the late nineteenth century with efforts to save specific species, such as waterfowl and bison, from overhunting. The movement led to the first wildlife refuges and national parks but also to the creation of the museum diorama. Over time, the conservation movement has undergone its own transformations as it adjusted to address new threats such as pesticides, air and water pollution, acid rain, and climate change. These developments are reflected in the changing nature of museum research, exhibitions, and education programs.

One of the primary objectives of the museum's programs has always been to improve the public's understanding and attitudes toward science and nature. When the museum was founded, nature was largely valued as a source of exploitable commodities by the state's growing business community and the settlers who were flooding into the region. There was little concern about future preservation. The Bell Museum has worked to change these attitudes and help build a broader affinity for nature and an appreciation for the multiple benefits of a healthy environment.

Like other institutions of science and conservation, the museum was long dominated by white males. This changed in the 1970s as more women at the museum earned graduate degrees and went on to hold important positions in museums, universities, and conservation agencies across the country. In the twenty-first century changes at the museum have continued; its current and previous directors have been women, and its new Breckenridge Chair of Ornithology, Sushma Reddy, is a woman of color. Still, much needs to be done to make the museum more inclusive of people from diverse backgrounds.

In particular, the museum has endeavored to be more inclusive in its public education and research roles. The traveling exhibition program developed in the 1980s included the exhibit *Many Faces of Science*, which showcased scientists and researchers of diverse backgrounds. The Honeybee Program initiated in the 1990s partnered racially isolated and non-isolated schools. In the 2000s, the museum's summer campers and counselors, as well as museum tour guides, have become much more reflective of Minnesota's current demographics. Current activities include seeking Artists in Residence of color, recruiting graduate students of underserved populations, and the research of museum curator George Weiblen in Papua New Guinea, which relies on the Indigenous para-taxonomists whose environment is being studied.

Throughout the Bell Museum's history there have been periods of rapid change and growth and periods of quiescence. There have been times when the different parts of the museum worked in harmony, and others when there has been competition and dispute. There have been times when the museum was the beneficiary of generous university and state support, and others when the institution was threatened by budget cuts and neglect. Through both good times and bad, the museum has managed to survive, innovate, and keep moving forward. During the dark days following a former governor's veto of its long-planned new building, the museum expanded its mission to include cosmology and planetary science by welcoming the Minnesota Planetarium, which was itself struggling to survive.

One of the constants in the museum's story is its connection to the study of birds. Thomas Sadler Roberts, the museum's longtime director in the early twentieth century, left his successful medical career to devote himself to the development of the museum because of his passion for birds. As director, he fostered the creation of the museum's early dioramas, initiated a Sunday afternoon lecture program illustrated with his own bird photographs, and made the museum a truly public institution. Arguably, his greatest achievement was the publication of his masterwork, *The Birds of Minnesota*. Illustrated by the leading bird artists of the day, the book put Minnesota on the ornithological map. It was compiled with the help of a wide range of amateur naturalists from around the state who faithfully sent Roberts reports of local bird life.

Thomas Sadler Roberts at the setting of the cornerstone for the museum's Church Street building in 1939.

The first meeting of the Minnesota Ornithologists' Union at the museum's new auditorium, 1940. University of Minnesota Archives.

Roberts's Minnesota correspondents, as well as students who had taken his ornithology class at the University, formed the nucleus for bird clubs that sprung up around the state in the 1920s and 1930s. After forming the Minnesota Ornithologists' Union (MOU) in 1938, the groups met again the following year in Duluth. At the meeting, Roberts invited the MOU to hold their next meeting in the auditorium of the new museum then under construction on the Minneapolis campus. This became an annual event, and the museum went on to supply the MOU with a mailing address and space to store the organization's records.

The building at 10 Church Street Southeast on the Minneapolis campus was the home of the museum for seventy-five years.

Native wildflowers bloom in the courtyard of the new Bell Museum on the University's St. Paul campus.

MOU members were welcomed to use the museum's library, and members would frequently bring in dead birds to add to the museum's collections. The MOU even helped sponsor big-name speakers at the museum, such as Roger Tory Peterson and David Sibley. In fact, many members have long thought of the Bell Museum as their "bird museum." When the museum, facing tough financial times, decided to raise funds for an endowed faculty position, ornithology was chosen as the area of expertise, and the MOU stepped up to help secure the funding for the Breckenridge Chair. From its earliest days, the museum's extraordinary strength in ornithology attracted top graduate students, many of whom went on to hold choice positions as curators and professors in museums and universities across the country.

From its humble roots as a one-room cabinet of curiosities at a one-building university in a frontier state, the Bell Museum did indeed grow, just as Newton Horace Winchell had hoped. In its new building on the St. Paul campus, it stands as a window into the research of what is now one of the world's great universities, while opening all of nature to children and expanding the public's appreciation for the natural world around us.

Participants from the Bdote Learning Center at the new museum's grand opening in 2018, honoring the Bell Museum's location on traditional and treaty land of the Dakota people. Photograph by Joe Szurszewski.

Old Main

Caribou diorama

BELL MUSEUM TIMELINE

1851 · University of Minnesota established, seven years before Minnesota's statehood

1872 · Geological and Natural History Survey is created by the Minnesota state legislature

1872–90 · Newton Horace Winchell, director of survey and first head of museum

1875–88 · Old Main is first home of General Museum

1889–1915 · Now known as the Zoological Museum, the museum moves to Pillsbury Hall

1889 · University establishes botany department

1890–1919 · Henry Nachtrieb, director of Zoological Museum

1890–93 · Menage expedition to the Philippines

1899–1902 · Research boat *Megalops* inventories aquatic life on Minnesota and Mississippi Rivers

1901–1907 · Josephine Tilden runs the Minnesota Seaside Station

1909 · James Ford Bell on collecting expedition for the Caribou diorama

1911 · Caribou diorama, the museum's first large diorama, is completed

1826–38 · John James Audubon produces *The Birds of America* double-elephant folio

1872 · Yellowstone is established as the nation's first national park

1891 · Itasca, Minnesota's first state park, is established

Newton Horace Winchell

Pillsbury Hall

Thomas Sadler Roberts

Church Street Museum

1915–19 · Thomas Sadler Roberts is associate curator of Zoological Museum and professor of ornithology

1916–39 · Zoology Building becomes the third home of the museum

1919–46 · Thomas Sadler Roberts, director of museum

1919 · Beaver diorama completed

1920s–40s · More than one hundred small portable dioramas are constructed for educational outreach

1926 · Walter J. Breckenridge begins work as a preparator at the museum

1928–38 · Name changed to Museum of Natural History

1928 · Audubon's *The Birds of America* double-elephant folio donated to the museum by W. O. Winston family

1932 · T. S. Roberts publishes *The Birds of Minnesota*

1938 · James Ford Bell and the Public Works Administration match funds to build a new museum

1939–67 · The museum is renamed Minnesota Museum of Natural History

1940–2017 · First building dedicated solely to the museum opens on Church Street on the Minneapolis campus

1940 · Wolves diorama, first large diorama painted by Francis Lee Jaques

1940–48 · Seven large dioramas created for the Church Street museum

1918 · Migratory Bird Treaty Act

1929 · Edwin Hubble discovers the universe is expanding

1934 · Federal Duck Stamp program begins

1934 · Roger Tory Peterson publishes his first field guide to birds

1939–45 · World War II · United States enters in 1941

Zoology Building

Wolves diorama

Viewing Wolves diorama, 1940s

Coniferous Forest (Cascade River) diorama

1946–69 · Walter J. Breckenridge, director of museum

1946 · John Jarosz hired as exhibits preparator

1947 · Minnesota state parks naturalist interpretive program started

1947 · Dwain Warner joins the museum as curator of ornithology

1947–48 · Museum visitors total 68,405, including 196 school groups

1948–49 · More than 18,000 visitors attend the Sunday lectures

1950 · Planetarium opens in downtown Minneapolis library

1954–56 · Cascade River (Coniferous Forest) diorama, the last large diorama, completed

1958–90 · Telemetry research begins at the University's Cedar Creek Natural History Area

1961 · Expanded planetarium opens in new Minneapolis Public Library

1964 · Passenger Pigeon diorama· Jaques paints his last diorama at Bell Museum

1964–99 · Frank McKinney is curator of ethology

1945 · DDT pesticide becomes commercially available

1953 · DNA double helix structure discovered

1957 · Sputnik, first satellite, launched

1962 · Rachel Carson's *Silent Spring* is published

Walter J. Breckenridge

Maxine Haarstick Planetarium Director

Passenger Pigeon diorama

Bud Tordoff and peregrine

Elephant skull in Touch and See Room

1965 · Addition to Church Street museum to house scientific collections and the Touch and See Room

1966–72 · Richard Barthelemy is first public education coordinator

1966 · Museum becomes part of the College of Biological Sciences

1967–2017 · Museum renamed James Ford Bell Museum of Natural History, after its longtime benefactor

1968 · Touch and See Room opens to the public

1970–83 · Harrison "Bud" Tordoff, director of museum

1970 · David Parmelee hired to direct Itasca Forestry and Biological Station and Cedar Creek Natural History Area

1972 · Jaques Gallery opened on mezzanine level with exhibition of artwork by Francis Lee Jaques

1982 · Midwest Peregrine Restoration Project started by Bud Tordoff and Patrick Redig

1982 · *Francis Lee Jaques: Artist–Naturalist*, museum's first traveling exhibition shown at the Smithsonian Institution

1983–90 · Donald Gilbertson, director of museum

1969 · Apollo 11: first humans land on the moon

1970 · First observation of Earth Day

1972 · DDT banned in the United States

1973 · Federal Endangered Species Act

1981 · Minnesota Endangered Species Act calls for state's first official list

Richard Barthelemy in the Touch and See Room

Upland Sandpiper by Vera Ming Wong

Family of Swans by Gary Moss

Honeybees on the roof

1984–2015 · *IMPRINT*, museum newsletter, is published for more than three decades

1987–2022 · Minnesota Biological Survey adds thousands of specimens to Bell Museum collections

1988 · *In the Realm of the Wild*, art by Bruno Liljefors, opens in newly remodeled West Gallery

1988 · *AIDS and Intimate Choices*, the museum's first traveling exhibit on a controversial issue

1990 · Museum becomes a regional site for the JASON Project's electronic field trips

1991–92 · Elmer Birney, director of museum

1993–94 · Kendall Corbin, director of museum

1993 · Scientific curators and their collections move to new Ecology Building, St. Paul campus

1993 · American Museum of Wildlife Art collection merges with Bell Museum

1993 · Robert Zink appointed as museum's first Breckenridge Chair of Ornithology

1995–2008 · Scott Lanyon, director of museum

1995 · Museum becomes part of the College of Natural Resources

1996 · University of Minnesota Herbarium joins the museum

1998–99 · JASON and Bell *LIVE!* reach more than 750,000 students through on-site and remote electronic connection

2002 · Museum's Honeybees and Pollinators educational program begins

2005 · *Minnesota, A History of the Land* premieres on Twin Cities Public Television

2006 · Museum becomes part of the College of Food, Agriculture and Natural Resource Sciences

1988 · Climate change recognized by scientists as serious threat

1990 · Hubble Space Telescope launched into Earth orbit

1990 · Human Genome Project started

2000 · Tree of Life Project launched by the National Science Foundation

IMPRINT, 1984

Sneezeweed, herbarium specimen

Carolina Parakeets by John James Audubon

New Bell Museum

2008–15 · Susan Weller, director of museum

2008 and 2009 · Governor uses line-item veto to cut funds for new museum building

2010 · Museum's Resident Artist Research Project begins

2011 · Minnesota Planetarium Society joins the Bell Museum

2014 · Museum exhibition, *Audubon and the Art of Birds*, has record attendance

2014 · State funding for new museum building approved, championed by State Representative Alice Hausman

2016 · Museum's Minnesota Biodiversity Atlas goes online

2016 · Earth Day groundbreaking celebration for new museum building

2016 · Denise Young becomes director of museum

2017 · Church Street museum closes after seventy-seven years

2018 · New Bell Museum, including 120-seat planetarium, opens on University's St. Paul campus

2018 · Name changes to Bell Museum

2018 · Sushma Reddy becomes Breckenridge Chair of Ornithology

2020 · Museum's scientific collections number more than 1.1 million specimens

2022 · Bell Museum celebrates its sesquicentennial, 1872–2022

2008 · Minnesota Clean Water, Land, and Legacy Amendment

2015 · First detection of gravitational waves

Susan Weller

Sushma Reddy with bird collections

Planetarium in new Bell

A Museum is Born

1872–1940

Documenting Minnesota

The Geological and Natural History Survey

Legacies are often born of singular events. In the case of the Bell Museum that event was the passage of the 1872 Geological and Natural History Survey Act. The legislation was enacted at a time when newly minted states were rushing to form statewide geologic surveys to determine the mineral holdings of lands appropriated from Indigenous peoples. Nonetheless, the act that created Minnesota's survey was unusual in several ways. First, the survey was housed in and run out of an academic institution—the University of Minnesota. A primary reason for this was to shield it from the political and economic interests that would inevitably try to exploit it.

Second, the legislation was written not by a politician but by an academic, William W. Folwell, the University's president. The act's language made it abundantly clear that the survey was to not only support the commercial exploitation of mineral resources but to also create a baseline for further scientific research on the state's natural resources and to provide the public with the fruits of that research. From Folwell's point of view, housing the survey within the University would also ensure that the school would be the center for all natural history collections and studies in the state.

To be sure, while the survey was to give primacy to "a complete account of the mineral kingdom as represented in the state," it was also to be "an examination of the vegetable productions of the state, embracing all trees, shrubs, herbs and grasses native or naturalized in the state [as well as] . . . a complete and scientific account of the animal kingdom as properly represented in the state, including all mammalia, fishes, reptiles, birds and insects." Further, a museum was to be established at the University to be the repository of all natural history and geological specimens collected.

Third, the act was unusual in specifying precisely what the state required of its natural history museum. The University's Board of Regents was "to cause proper specimens, skillfully prepared, secured and labelled of all rocks, soils, ores, coals, fossils, cements, building stones, plants, woods, skins and skeletons of animals, birds, insects and fishes, and other mineral, vegetable and animal substances and organisms discovered or examined in the course of said surveys."

Fourth, written into the act was the stipulation that an effort be made to prepare duplicates of specimens collected "for the purpose of exchanges with other state universities and scientific institutions, of which latter the Smithsonian Institute at Washington shall have the preference."

In a move that further burnished the survey's credibility, Folwell hired Newton Horace Winchell to head the survey and the museum. Winchell was a veteran of Michigan's state survey and also carried a famous name. He was the younger brother of nationally respected geologist Alexander Winchell. In Newton Winchell, the University acquired not only a first-class scientist but an all-around scholar whose professional interests extended beyond simply geology. While at the University, he taught not only geology but botany and zoology as well. Winchell also proved to be an able administrator with a knack for getting his way.

Newton Winchell, a geologist, was the first director of the survey. University of Minnesota Archives.

His most immediate challenge upon assuming his position was to do something about the pittance (a mere $1,000 per annum)

Natural History Survey crew at Long Lake, Minnesota, 1893. Left to right: Professor Oestlund, Professor Lee, Henry Nachtrieb, Conway MacMillan, A. P. Anderson, Josephine Tilden, a student from Upsala, and an unknown botanist. University of Minnesota Archives.

the legislature had allocated to the survey to do its work. Winchell had a better idea, and the year after he took over, he convinced the legislature to take acreage that had been set aside by the state specifically for development of salt deposits that were thought to be present in the state, and to have title to that acreage transferred to the University. The sale of those "salt lands," as they were called, would now be used to support the work of the survey. At the legislature's behest, Winchell himself drafted the legislation. The legislature quickly passed the plan and upped its annual appropriation to the survey to $2,000 until the "salt lands" became available.

Two thousand dollars was still not enough, of course, and Winchell was forced to take on side jobs to get by. In 1874, an opportunity arose for him to join General George Armstrong Custer's expedition to the Black Hills, serving as geologist and botanist. Although he couldn't have known it, Winchell's presence on this trip would involve the museum—however tangentially—in one of the nation's darkest moments. Custer's expedition not only violated the 1868 Treaty of Fort Laramie, it launched a Black Hills gold rush and a wave of subsequent treaty violations that would culminate in the 1890 massacre of Native people at Wounded Knee.

Known simply as the "general museum," the space Winchell had chosen for the facility was located on the third floor of Old Main, the principal building on the University's mostly empty campus grounds. One room was reserved for his geologic specimens, and another for the zoological and botanical collections, which now included selected fruits of Winchell's time with the Custer expedition: three stuffed pronghorn antelopes, two white-tailed deer, an elk, a grizzly bear, a mule deer, a badger, a moose collected in Minnesota, and eventually even Custer's dog.

Winchell had a well-defined vision for the museum. In his report to the 1877 legislature, he laid out what he saw as its future: "If it grows and is properly cared for, as the state grows, it will become the pride not only of the University, but of the whole state, and will delight and instruct the hundreds of students who resort there for instruction, and the citizens of the state and others who may desire to visit it."

In 1875, Winchell kicked the survey into high gear and began filling the museum in earnest. He took on teenager Clarence Herrick as his lab and museum assistant. Herrick, a dedicated naturalist, was a close high school friend of Thomas Sadler Roberts, who would later become the museum's third director. In fact, Roberts joined the survey four years later to collect along the wild north shore of Lake Superior. Indeed, in those early years of the survey, between 1875 and 1880, the lion's share of specimens coming into the museum were collected by Herrick and Roberts.

The zoological alcove in the General Museum as it was housed in the University's Old Main building. University of Minnesota Archives.

Henry Nachtrieb in his laboratory at the University. University of Minnesota Archives.

Winchell also began a comprehensive botanical survey of the state in 1875. Botanists around the state were asked to identify and describe all species in their regions. Field workers were asked to organize data according to a strict set of guidelines. They were also urged to preserve as many specimens as possible. Classification was to follow reference books that Winchell recommended.

In 1888, with the workload and collections growing, Winchell convinced the University to create the position of state zoologist to oversee the survey's zoological investigations and to head the museum. To fill the post, which now also supervised a state ornithologist and state entomologist, the Board of Regents chose Henry Nachtrieb, a former student of Winchell, who had already been hired to teach. Then, in 1890, at Winchell's urging, the survey was divided into three lines of activities: geology, headed by Winchell; zoology, headed by Nachtrieb; and botany, to be headed by Conway MacMillan, the University's first botany professor, who had begun collecting for the survey in 1887.

Nachtrieb proved an able leader, initiating a set of mussel and fish surveys of select rivers of the state. In 1899, he launched a first-ever major study of the fish in the southern Minnesota watersheds of the Minnesota and Mississippi Rivers. In fact, it was a literal launch, for the field station he had designed was a floating lab—basically a shack on a floating platform. It had enough space to accommodate lab work, living quarters, and a tiny kitchen. Naming the vessel *Megalops* after a larval stage of the blue crab, the species that had been the subject of his graduate research, Nachtrieb sent the craft down the two waterways and up their main tributaries, collecting a trove of early biological information. Survey staff at various times and locations included entomologists, herpetologists, ichthyologists, limnologists, and other specialists as needed.

Entering the new century, Winchell had even bigger plans for the survey: opening another museum to house the mineral and rock collection as well as offices for graduate students. These dreams, however, flew in the face of the economic and institutional realities. The survey had sold most of its source of annual funding, the lands set aside for unsuccessful salt exploration. Absent another source of funding, there was no way the survey could be sustained, and the Board of Regents was not interested in finding one. Other departments within the University were now entirely capable of doing the survey's work.

Winchell pushed hard, and in the confrontation that followed he was fired. With Winchell's exit in 1900, the Board of Regents decided to discontinue the survey's geological inquiries, transferring them to the department of geology. The botanical and zoological studies were continued as part of the survey under the direction of Nachtrieb and MacMillan, but not for long. By 1903 the survey's funds had run out. Nachtrieb continued as the museum's director until 1919, but for all practical purposes the survey as a separate enterprise was done. In 1916, its last vestiges effectively disappeared when its zoological activities were transferred to the museum and what remained of its botanical collections went to the herbarium in the University's botany department.

The survey's geological work was reinstated later but only briefly. Then it, too, was discontinued. The survey may be gone, but its legacy lives on. Today, its successors can be found in the ongoing Minnesota Geological Survey, the Minnesota Biological Survey, and, of course, the Bell Museum.

The Megalops, Nachtrieb's floating field lab, anchored in the Mississippi River north of Brownsville, Minnesota, circa 1902.

Henry Nachtrieb conducts a fish survey at Lake Pepin in 1900. The group netted one thousand pounds of paddlefish, a species now listed as threatened in Minnesota.

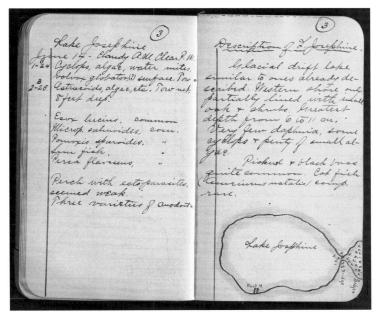

Pages from Nachtrieb's Zoological Survey field notebook, 1892. University of Minnesota Archives.

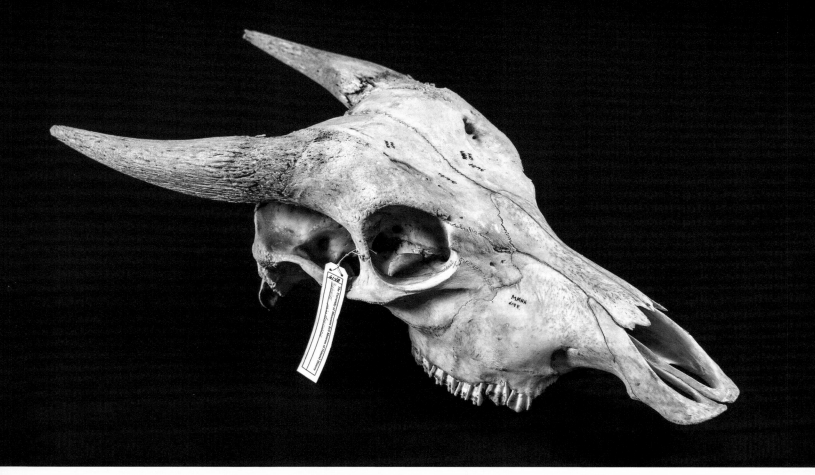

Many of the species collected by the Menage expedition are now rare. The tamaraw, or Mindoro dwarf buffalo, is the only bovine native to the Philippines. Human population growth, habitat loss, hunting, and logging have drastically reduced their numbers, and they are now considered critically endangered. Photograph by Joe Szurszewski.

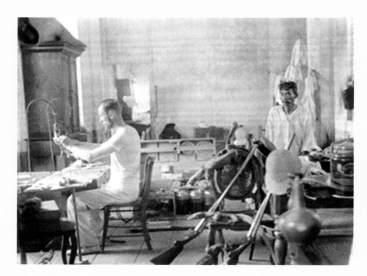

The workroom for the Menage expedition at Capiz, Panay, Philippines, October 1890. University of Michigan Library, Special Collections Research Center.

Dean Worcester and Frank Bourns with a local guide in the Philippine Islands. Hennepin County Library.

The Menage Expedition

How Orangutan Bones Landed in the Bell Museum Collections

So what do moonrats, stink badgers, giant flying foxes, dwarf buffaloes, sunbirds, and barking deer have in common? For one thing, they are animals native to the Philippine Islands. For another, they have all found a most unlikely home in the Bell Museum's collections.

The story begins sometime around 1890 when Horace Winchell, son of Newton Winchell (then head of the Minnesota Geological and Natural History Survey), agreed to help two of his college buddies, Dean Worcester and Frank Bourns, scientists with the University of Michigan. The two were seeking institutional backing for a collecting expedition to the then little-studied Philippines. They were already familiar with the islands, having been part of an earlier collecting expedition. Horace pitched the idea to the Minnesota Academy of Natural Sciences, an organization cofounded by his father. The academy not only agreed to lend its support but one of its new members, Louis F. Menage, a wealthy Minneapolis financier and real estate developer, offered to pay for it. Worcester and Bourns would organize the trip and do the collecting, and the academy would curate and care for whatever they brought back. Menage would supply $14,000 to pay for it all; in return, he asked that the expedition be named after him.

Beginning in 1890, and over the next twenty-nine months, Worcester and Bourns visited nineteen islands, traveling by steamers, freighters, and even dugout canoes. Using local men and boys as field hands, they hunted, trapped, and snared whatever they could find. Sometimes they bartered species they had collected for species they felt they needed. Sometimes they were able to purchase whole collections of such things as butterflies and mollusks. In the end they managed to bring back some four thousand bird specimens and over five hundred mammals, in addition to numerous reptiles, fish, and mollusks.

Upon their return in 1893, the two scientists prepared to settle down to the work of cataloging and describing their specimens. Unfortunately for them and the Minnesota Academy of Natural Sciences, their return coincided with the onset of what would be known as the Financial Panic of 1893. The panic not only devastated Menage financially, it exposed a $4 million fraudulent real estate loan scheme that he was running. To escape prosecution, Menage fled to South America, and the academy was left to deal with the fallout. Deeply embarrassed and with the spoils of their project now tainted by ill-gotten money, academy members were unable to raise the additional funds needed to do the postexpedition work of cataloging and curating the collected specimens. The collection was subsequently dispersed. Some specimens were given to the University of Minnesota, some to Hamline University, and some were sold to the Field Museum of Natural History in Chicago. A few were shipped to institutions in London and Dublin. The rest were retained by the Academy of Natural Sciences, which later gave them to the Minneapolis Public Library when the academy disbanded. These specimens were subsequently given to the Bell Museum in the 1950s.

Although these bones, shells, skins, and other remains are old, they are still highly valuable to science and conservation today. The habitats from which they came have been practically wiped out, but these relics from the past give us an idea of the biological diversity those ecosystems once held and how they might have functioned—vital information for any effort to restore these special places.

A marine alga, Egregia menziesii, *collected by Josephine Tilden in 1898.*

Josephine Tilden

Paving the Way for Women in Science

Natural history museum collections are quite literally libraries of life. These specimens—even the humblest of them—have stories to tell, not only about themselves and the environments they came from but often about the people who collected them. One such tale lies in a humble collection of algae and seaweed kept in the Bell Museum's herbarium. The story these dried plants tell is of Josephine Tilden, the University of Minnesota's first female staff scientist. It is a narrative that could have jumped from the pages of a conventional novel: a strong-willed and determined woman lived a life of physical adventure, overcoming great odds to become a world-renowned scientist, but her controversial ways ultimately antagonized the University administration and led to a bitter feud with the leadership of her own department that lasted until she retired.

The tale begins in 1890 when the twenty-one-year-old Tilden, a graduate of Minneapolis Central High School, walked into the classroom of Conway MacMillan, the head of the University's botany department. By all accounts MacMillan was a formidable presence who did not suffer fools gladly. He was a respected scientist but also one who did not mind breaking with convention. Maybe that is why he took a decided interest in promoting Tilden's career. Tilden was one of a small but growing number of women nationwide who were trying to break into the ranks of the male-dominated sciences. MacMillan became her mentor and champion.

Josephine Tilden in 1924. University of Minnesota Archives.

Under MacMillan's tutelage Tilden earned her bachelor of science in botany from the University in 1895, a master's degree two years later, and then became a department instructor. Along the way she had become particularly interested in phycology (the study of algae and seaweed), a line of research both MacMillan and University president Cyrus Northrop encouraged her to pursue.

Tilden was interested in the little-studied algae of the Pacific coast. With MacMillan's backing she convinced Northrop that despite the distance and expense of conducting such research, her work there would be worth it. Northrop not only supported the project but promised her that if she committed herself to the study for five years, she would be made a full professor. It was a commitment she exceeded by a lifetime.

In 1898, the twenty-nine-year-old Tilden and her mother, Elizabeth (a young woman did not travel alone in those days), headed west by train to Vancouver, Canada. On August 4, in driving rain, the two women set out in a rowboat from the small community of Port Renfrew on the west coast of Vancouver Island and headed north through the rough waters of the Juan de Fuca Strait, bound for an isolated beach a few miles up the coast and sixty-five miles west of the city of Vancouver. Their guide and rower was Thomas Baird, the man who owned the beach and much of the land around it. They arrived at low tide to discover just what Tilden was looking for: a stretch of sandstone shoreline pitted with countless tidal pools crammed with more species of algae than she had ever seen in one place. "The

A class works on the rocks during low tide near the Minnesota Seaside Station on Vancouver Island, 1901–06. Photograph by Ned Huff. University of Minnesota Archives.

algae covering that exposed shore were beyond my wildest dreams," she wrote later.

Living on beans and tea, the three spent the next four days, much of it in the rain, gathering specimens. Recalled Tilden, "At the end of the fourth day, Mr. Baird said to me, 'I am going to give you a deed for the best four acres on my place. Make your choice.'" The spot she chose—now known as Botanical Beach—was the very place where she had landed. Here, in 1901, she founded a research facility she called the Minnesota Seaside Station.

The University agreed to supply instructors and equipment but no money, leaving Tilden and MacMillan, who had joined her in operating the station, to run the facility and its programs with their personal funds. Over the next six years, the station became an environmental center of sorts, offering courses in geology, zoology, algology, and lichenology. Every summer, twenty-five to thirty students—most of them women—would trek there from schools throughout the Midwest to do research and attend lectures. After the sixth year, Tilden and MacMillan offered to donate the land and buildings to the University if the school would take over its administration. The Board of Regents refused, citing their concern that it was inappropriate for the University to own land in a foreign country.

Angered and frustrated by the decision but unable to continue maintaining the facility with their own funds, the two had to close down the Seaside Station after the 1907 season. Both suspected that the University was really more interested in pursuing other, less exotic (and less costly) research closer to home. Indeed, the University opened the Forestry and Biological Station at Lake Itasca a couple of years later. MacMillan angrily resigned from the University, but Tilden, now an assistant professor of botany, stayed on the University payroll, becoming a full professor in 1910.

Tilden went on to become a leading international authority on the phycology of the coastal waters of the Pacific, but her relationship with the University and the chair of the botany department, C. O. Rosendahl, was increasingly

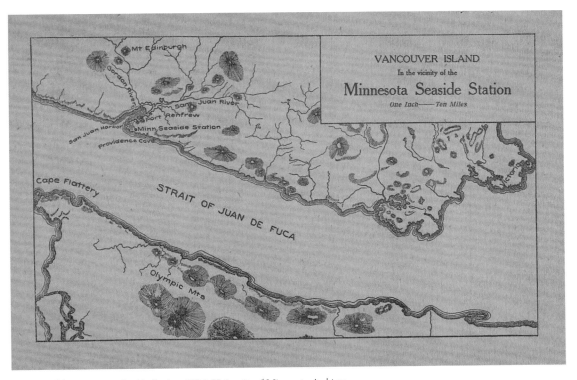

Location of the Minnesota Seaside Station, 1906. University of Minnesota Archives.

strained. Her determination to make the University a significant center for Pacific algal research, her independent ways, and the costs of her expeditions—thirteen of them to destinations all around the Pacific rim—rankled Rosendahl and the administration. Her relationship with Rosendahl became all but irreconcilable when Tilden put together an unsanctioned seven-month collecting trip around the world with ten students. Tilden raised money from the students, from James Ford Bell, who supplied about $5,000 (equivalent to about $87,000 today), and from contributed funds of her own. She was supposed to repay the funds from the sale of some of the collected specimens, but that never happened. Her group generated fifty-six trunks of dried plants, which were shipped home, and Tilden charged the shipping costs and equipment purchased directly to the botany department.

When Tilden retired in 1937, Rosendahl, according to department documents, hired a guard at the herbarium to prevent her from removing any of her collection. But in typical Tilden fashion, she got around this move by persuading the Board of Regents to loan her the collection. The board agreed, and more than one thousand cartons of her specimens, books, and notes were sent to Florida, where Tilden had retired.

When Tilden died in 1957, she bequeathed her collection to Joseph Wachter, a friend who owned an organic seafood company in San Francisco. The Organic Sea Products Company now houses a Josephine Tilden Museum and Library that contains many of her notes and books. Wachter sent her collection, three hundred boxes of algae specimens, back to the University. By 1967 marine algae was no longer a research topic at the University, so a new home was sought for Tilden's collection. No takers were found. Today it remains a part of the University of Minnesota Herbarium at the Bell Museum. The collection continues to offer important historical data to researchers near and far. In recent years, with the growing industrial use of algae in bioproducts and biotechnology, Tilden's collection has been of renewed interest. Josephine Tilden, who had long been intrigued in the economic uses of algae, would no doubt be pleased.

Students climb on rocks after being caught by the rising tide. Photograph by Ned Huff. University of Minnesota Archives.

Student collecting starfish and other zoological specimens during low tide, 1901–06. Photograph by Ned Huff. University of Minnesota Archives.

Thomas Sadler Roberts

Naturalist, Doctor, Director

Physician, ornithologist, author, photographer, professor, and pillar of his community—Thomas Sadler Roberts was all of these. But perhaps his most impressive accomplishment was what he did with a collection of natural history specimens tucked away in the bowels of the University of Minnesota, turning a menagerie of plants and stuffed animals into a first-rate natural history museum.

The Roberts family moved to Minneapolis in 1867, when young Thomas was nine years old. The move from Philadelphia was prompted by the hope that wide-open landscapes would help restore his father's health. The elder Roberts suffered from "consumption" (now known as tuberculosis), and his prescription of fresh air and sunshine became a perfect excuse for daily outings, frequently with his two sons, Thomas and John Walter. Traveling by foot or by horse and buggy, even short excursions in those days were often all-day affairs. As the trio traversed the outlying oak savannas, tallgrass prairies, and maple-basswood woodlands, they became witnesses to the state's abundant diversity.

On one such outing in the spring of 1874, on an open prairie just three miles from the center of town, they encountered "great flocks" of migrating American golden plovers, which a much older T. S. Roberts would later compare to those of the passenger pigeon. On another occasion, they witnessed a huge flock of brilliantly clad yellow-rumped warblers. The sight aroused in young Roberts "feelings of wonder and delight . . . that have never been repeated." Later that summer they observed "large flocks" of passenger pigeons in the oak woodlands along the shoreline of Lake Johanna, swallow-tailed kites "soaring continually" near Buffalo Lake in

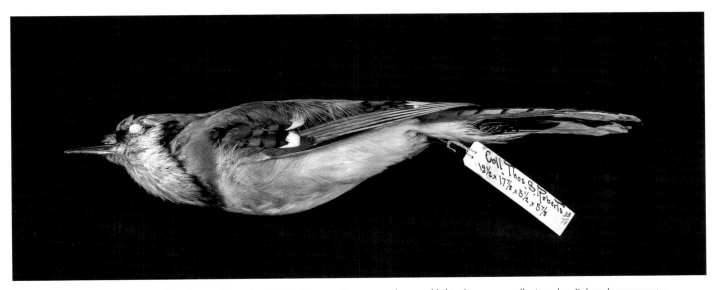

This blue jay, collected by Thomas Sadler Roberts in December 1876 in Minneapolis, was one of many added to the museum collection when Roberts became curator. Photograph by Joe Szurszewski.

Topographic survey party, 1881. Thomas Sadler Roberts (lower left) with John Cobb, Nathan Butler, Henry Nachtrieb, and Edison Gaylord. University of Minnesota Archives.

Wright County, and upland plovers, which were "one of the most common and noticeable birds in the open country about Minneapolis." Roberts relished the diversity and sheer abundance of wildlife, especially birds. These experiences became the foundation of his evolving conservation ethic and his dedication to public education. They also infected him with a deep desire to create a comprehensive guide to all the native bird life of the state—their names, their physical descriptions, their habits, their vocalizations, their migration and nesting behavior—a dream that would occupy him for much of his adult life.

When the budding ornithologist was a freshman in high school, he, along with his best friend, Clarence Herrick, and three other classmates, formed the Young Naturalists' Society. The boys spent much of their free time collecting plants, shooting and trapping animals for study, and identifying and documenting species. They met regularly to discuss their research and scientific issues of the day. The society assembled an impressive collection of plants, bird skins, and research papers, much of which would eventually become the property of the Minnesota Geological and Natural History Survey.

The survey, enacted in 1872 by the state legislature, was headed by University professor Newton Horace Winchell, a geologist by training. When Clarence Herrick went to work for Winchell the next year, Roberts and the Young Naturalists' Society suddenly had a direct pipeline to the professional scientists who were doing the real exploring, pushing at boundaries of the then-known natural world.

In 1878, Roberts enrolled at the University of Minnesota. That summer he joined the survey, exploring the wilds of the roadless north shore of Lake Superior. He then spent several summers as a land examiner for James J. Hill's St. Paul, Minneapolis & Manitoba Railway, traveling the state and getting to know its widely disparate regions and landscapes. In this job he became friends with fellow surveyor Henry Nachtrieb.

At the same time, Roberts had begun submitting his bird notes to prominent ornithological and popular outdoor publications. He started corresponding with noted figures in ornithology, such as conservationist George Bird Grinnell (editor of *Forest and Stream*); William Brewster, a founding member of the Nuttall Ornithological Club of Cambridge, Massachusetts; the noted ornithologist Elliott Coues; and Robert Ridgway of the Smithsonian Institution, a founder of the American Ornithologists' Union.

In 1881, Roberts reentered school, enrolling in the University of Pennsylvania's medical school. Five years later he returned to Minneapolis, a newly minted doctor. He opened a practice (at first largely based on former friends and his father's wealthy contacts) and married his hometown sweetheart, Jennie Cleveland.

The relationships, both professional and personal, that Roberts made during this time became critically important for the career he would segue into, for these same families would provide much of the financial support that enabled Roberts to build the University's natural history museum into a

Leslie Dart and Roberts set out to photograph birds at Long Meadow Marsh, 1900.

Jennie Roberts outside her husband's medical office. University of Minnesota Archives.

first-class institution. Significant among them was the family of flour-milling magnate James Stroud Bell. Roberts especially hit it off with Bell's son Jim. Like Roberts, young James Ford Bell had a passion for birds and for the natural world.

A DOUBLE LIFE

The next twenty years were among the most frenetic of Roberts's career as he essentially led a double life pursuing both his medical practice and his private passion of ornithology. Sometimes carving weeks out of his demanding medical practice, he would take extensive field trips documenting, collecting, and photographing birds. Memorable trips were taken to Heron Lake, an expansive wetland in southwestern Minnesota that supported thousands of migrant and nesting birds, including Franklin's gulls, black-crowned night-herons, redheads, and canvasbacks. There he spent time with local market hunter Tom Miller, who was "surprisingly well-informed" regarding "the intricate pathways and teeming bird-life of that vast wilderness of reeds and rushes." Other outings with colleagues included trips along the Mississippi River's backwaters, where they spent time tracking down the elusive prothonotary warbler, and journeys documenting the dazzling diversity of warblers in the boreal forests surrounding Lake Vermilion.

With his medical workload piling up, in 1898 Roberts hired Mabel Densmore as a full-time assistant and William Kilgore as a clerk to help him with his bird projects. Leslie Dart, a friend and senior intern at Minneapolis's St. Barnabas Hospital, where Roberts was chief of staff, also pitched in to assist Roberts in his birding pursuits. Indeed, Dart became a kind of de facto medical partner, tending Roberts's patients when Roberts was in the field. During this time Roberts began to get serious about writing his definitive book on Minnesota birds. He began assembling a small army of trusted bird observers across the state who would regularly send him field observations that he would compile into a growing database of information.

His double life was not to last. As his obstetrics practice grew, so did the demands of his caseload. In 1901 Roberts was appointed professor of pediatrics at the University of Minnesota, and in 1906 he was made clinical professor of

Roberts botanizes in a birch grove. University of Minnesota Archives.

Mabel Densmore, Roberts's indispensable assistant both in his medical practice and with the publication of The Birds of Minnesota. University of Minnesota Archives.

children's diseases. His field excursions came to an end.

Not until the senior Bell died in 1915 did Roberts feel he could close his practice without feeling he was abandoning his longtime patients. Other factors also appeared to have played a role. Primary among these was the fact that the Natural History Survey's fledgling museum was about to move into a new, more spacious building. Roberts now felt financially secure enough to take on a new career for which he was not likely to be paid. In any event, he joined his friend and former colleague Henry Nachtrieb, who had assumed the directorship of the Natural History Survey. Roberts was given the title of associate curator of the Zoological Museum and professor of ornithology. His salary? Nothing.

The following year, the entire zoology department and museum moved out of Pillsbury Hall into the new Animal Biology Building. Roberts set about establishing the museum he had long dreamed of—one organized around the state's major biomes, depicting plants and animals in their native habitats, a museum that welcomed the public. His first project was the Beaver diorama.

In 1919, Roberts assumed a new title, and the museum a new name. He was now the director of the University's Zoological Museum. That same year, he started teaching a course in ornithology, using the city's streetcars to take students on local field trips, and he began traveling to small community libraries and schools around the state with his self-produced "Home Life of Birds" film. Building on his commitment to create a public museum, he also launched a publicity campaign to encourage visitors to come to the facility. One early successful program was a series of natural history lectures for families, held on Sunday afternoons from January through March. Invited speakers came from University natural science departments and state agencies. A main attraction of these talks was movies of animals in the wild, thanks to

Roberts works with the bird collection, 1917.

Jim Bell, who had purchased a projector for the museum. The footage shown often included Roberts's own, from his early experimentation with moving picture technology.

The 1920s was a watershed decade at the museum as the Heron Lake and the Bears dioramas were completed and work was started on the Pipestone Prairie display. As Roberts's workload increased, he hired William Kilgore as a full-time assistant to work with students to prepare skins, lead tours of the museum, and assist Roberts on his collecting trips. In 1926, when Jenness Richardson, the museum preparator left, Roberts hired Walter Breckenridge, freshly graduated from the University of Iowa's museum program. It was a pivotal hire, as Breckenridge would go on to succeed Roberts as museum director.

By then, much of the natural world that had so greatly shaped Roberts's worldview was rapidly disappearing. In 1924, on a collecting trip to far western Minnesota to revisit an area he had last seen forty-five years before, the full extent of the loss became apparent to him. Where once there had been boundless prairie, there were now mostly agricultural fields devoid of the plants and animals he had so marveled at.

Now in his sixties, Roberts frequently reflected on the enormous changes he had witnessed during his lifetime. As the state changed "from a comparative wilderness to a populous and prosperous commonwealth," forests had been leveled, marshes and shallow lakes had been drained, and streams had become polluted. Committed to promoting a better understanding of environmental conditions and natural history, Roberts believed that public education was critical so that citizens "will better understand the problems involved in the unbalance created by man, and will seek intelligently for such remedies as may be possible."

Galvanized by these changes and the passing of time, Roberts made his first priority assembling his treatise on

University bird class taught by Roberts in 1919. Olga Lakela, who would become a botany professor at the University of Minnesota, Duluth, is at upper left. University of Minnesota Archives.

Minnesota birds. He rehired Mabel Densmore to begin compiling the nearly sixty years' worth of field notes that he had amassed, most of them in longhand. These were not only Roberts's notes but also those sent in by his far-flung network of 160-plus correspondents. For the next seven years she laboriously organized these documents and typed them, becoming his indispensable partner in the endeavor. On June 6, 1932, the two-volume *Birds of Minnesota* went on sale. Within three days, the entire first run of a thousand books had sold out. A thousand more were printed and were gone by mid-July.

That year Roberts's wife, Jennie, died. Despite this profound loss, Roberts plowed on. Now deep into the Great Depression, the nation was in the grip of prolonged drought. Wildlife populations were plummeting, and game management had become a recognized discipline and profession. Roberts strode right into the thick of the hottest conservation issues of the time. Advocating for closed seasons on hunted species and the aggressive use of refuges to protect and rebuild wildlife populations, he found himself going head-to-head with the conventional wisdom of the day that simply blamed predators (hawks, owls, etc.) for loss of game. In articles, on the radio, and on the lecture circuit, Roberts was one of the early voices decrying habitat loss and the impact of agricultural drainage. "Sportsmen," he wrote, "will have to become less eager, less bent upon killing and will have to view the situation more broad-mindedly. They will unquestionably be obliged to accept even greater restrictions."

In 1938, Roberts marked his seventy-eighth year by joining forces with James Ford Bell to convince the Regents of the University to submit a proposal to the Public Works Administration (PWA) for $122,000 to build a new natural history museum. Ensconced as they were in the old Zoology Building (formerly called the Animal Biology Building), the museum's collections had once again outgrown their home.

Roberts with a passenger pigeon specimen and his book The Birds of Minnesota, *published in 1932. University of Minnesota Archives.*

Bell promised to match the government funds with $150,000 of his own. That fall, ground was broken for the new building at the corner of Church Street and University Avenue. That same year, Roberts was awarded the prestigious Brewster Medal, the profession's highest honor, for his contributions to ornithology.

Accolades aside, there was still more to be done. Roberts's dream of a new museum had come to fruition, opening for business in 1940. Now it was time to fill it. One of the first actions he took was to establish a natural history reference library within the museum, initially stocking it with the entirety of his own vast collection of books and documents. Over 1,000 bound volumes and more than 4,200 other related documents became the property of the University that next year. Then, with the critical and timely help of James Ford Bell, Roberts got a reluctant American Museum of Natural History to "loan" him Francis Lee Jaques, a Minnesota native and the best-known diorama painter of the day. Between 1940 and 1945, Jaques finished backdrops for the Wolves, Lake Pepin, Snow Geese, and Sandhill Cranes dioramas.

Roberts was now pushing eighty-seven. He had managed to come through a series of small strokes that had weakened him considerably. Amazingly, he was still teaching his bird classes and tending to the ills of a small coterie of patients, longtime friends. And he was still pulling in the money for his beloved museum. In July 1945, the MacMillan family (patients of his) agreed to fund construction of an elk diorama. Roberts, lamentably, never lived to see it. In failing health, he died on April 19, 1946, at age eighty-eight.

The caribou group was the first diorama created at the museum. It was moved from Pillsbury Hall to the Zoology Building in 1919, then into the museum's Church Street building in 1940.

Making a Museum for the Public
The Early Dioramas

It was cold that early November day in 1909 when James Ford Bell decided he'd had enough. The young Minnesota businessman was leading a collecting expedition to the barren interior of Newfoundland to hunt woodland caribou, and the party's prospects were not looking good. The group had been in the field for nearly a month. The weather had become foul—high winds, driving rain, sleet—and a blizzard was yet to come. They had paddled up rivers, portaged around falls, and waded through icy rapids and soggy bogs. Their main boat and one of the canoes had been bashed on rocks. Their supplies were nearly gone. And now the expedition, too, appeared to be on the rocks. Clearly, it was time to go home.

But this was no wealthy man's recreational outing, and Bell had invested too much of himself in its success to call it quits. He decided to press on.

A leader in the grain milling industry, an avid outdoorsman, and a historian of exploration, James Ford Bell was also a committed conservationist. He had brought with him on this trip Frederick Atkinson, a business associate, and the taxidermist Charles Brandler. The expedition was guided by a group of local Mi'kmaq (Micmac) Indians. The landscape was desolate—barren, burned-over terrain covered with fallen trees and muskeg. The caribou were not as plentiful as they had expected. Atkinson was out to get a trophy male and had shot several large bulls in the hope of getting one with a massive rack and perfectly formed head that would go well on his wall. Bell was interested in something quite different.

As well as hunting, Bell was taking photos of the surrounding landscape and the caribou herds. Brandler, who worked at the Milwaukee Public Museum and later for the Field Museum in Chicago, was not just taking the heads and antlers but skinning out entire hides, making plaster casts of key parts, and saving the bones. He would need all of them in order to mount the animals in lifelike poses in a new type of exhibit called "habitat groups." Bell had seen these re-creations of wildlife in their natural habitats at other museums and wanted to bring this new concept to Minnesota.

Up until then, the small museum at the University of Minnesota consisted of collections of rocks and minerals, bird skins, skeletons, and fish preserved in jars of alcohol. To this, Bell now wished to add his caribou scene. Completed in 1911, the exhibit featured a mature bull caribou facing off with a younger male. A female and yearling are nearby. They stand in hummocks of sphagnum moss. A painted landscape of open muskeg with stunted black spruce and tamarack forms the background. Dark clouds and squalls of cold rain sweep across the sky. The exhibit—a habitat diorama—is a virtual window into nature, where the animals are shown living in their natural habitat. The diorama creates an illusion that transports the visitor to a location, and a story they may never experience in real life.

ORIGINS OF THE DIORAMA

The natural history museum has roots dating back to the Renaissance. The first museums were literally curiosity cabinets: collections of the rare and the fabulous—alleged unicorn horns, giant bones, thunderbolts, and claws of the mythical griffin. In the sixteenth and seventeenth centuries and in the advent of the Age of Reason, these exotic assemblages evolved into private scholarly collections. Naturalists of the day replaced fantastic items with a broad array of representative specimens from the major groups of plants and animals.

Not until the eighteenth century did the first exhibits for the general public begin to appear. The artist and naturalist Charles Wilson Peale opened a museum in Philadelphia in 1786. Peale experimented with new techniques to preserve animals in lifelike poses and even included painted landscape backdrops. Tragically these early attempts to integrate art and science for public education were lost in a devastating fire.

Through much of the nineteenth century, natural history museums strove to be more scientific with displays based on taxonomic classification. Museum galleries were filled with glass cases with row upon row of specimens grouped by their relatedness. These displays were of interest to collectors, while the general public seemed more interested in the freakish and melodramatic displays of wildlife that populated circus sideshows, world fairs, and especially P. T. Barnum's exhibitions.

But the American view of nature was evolving. With the growth of Romanticism in the early nineteenth century, the solitude and mystery of wilderness began to be revered. This attitude was expressed in the works of such American landscape painters and artistic masters as Frederic Church and Thomas Moran. By the late nineteenth century, that actual, physical landscape was rapidly changing. Native people were being relocated and contained. Settlement and overhunting were driving wildlife, such as bison, passenger pigeons, and some waterfowl, toward extinction. Museums began to see themselves not just as repositories for scientific collections but as places for popular education. Museum educators wished to promote the fledgling conservation movement and protect what they considered to be a vanishing wilderness. They combined the new science of ecology and advances in taxidermy with the art of American landscape painting to produce exhibits that depicted the interrelationships of animals, plants, and the environment. This new exhibit form marked the debut of the modern natural history diorama.

DEVELOPMENT OF THE BELL MUSEUM DIORAMAS

Throughout its history, the Bell Museum has mirrored the key stages in the evolution of natural history exhibits. Among the collections in the early museum were such novelties as mounted antelope, elk, grizzly bear, and badger. When Thomas Sadler Roberts became director of the museum in 1919, it began a dramatic transformation. Referring to the old collections as "a heterogeneous assemblage of mounted mammals," Roberts stated:

> The old style of zoological museum, thank goodness, is gone forever. It has happily disappeared with the old style of taxidermy. Single specimens, clumsily stuffed in ghastly contrast to life and standing in an apathetic row, do not meet modern ideas of teaching zoology. Nor is the modern museum intended solely for students—it is for everybody to see and enjoy and learn about the interesting wild things that are disappearing or getting too remote to be easily found. (*Annals of Museum of Natural History*, T. S. Roberts, 1939)

Roberts threw out the "pathetic and moth-eaten mockeries of Minnesota animals" and embarked on the creation of a modern museum.

The oldest of the Bell Museum's dioramas, the Caribou, Dall Sheep, and White-tailed Deer, were completed in the 1910s when the art of diorama making was still in its infancy. They featured big-game animals in natural settings. Though a definite advance over trophy mounts or sideshow displays, they often mistakenly showed the species in unnatural family

Jenness Richardson casts the tail of a beaver in plaster so that he can reproduce it for the Beaver diorama, 1917.

Wax leaves and plants created for the Beaver diorama.

groups and did not depict much about the animal's ecological setting. The Beaver diorama, the first completed under Roberts's direction in 1919, was quite different. He hoped the new museum would develop in future generations a "greatly increased number of intelligent and efficient conservationists of our natural resources." His vision for the dioramas was to "have an esthetic appeal akin to that of the best displays in our art galleries."

Roberts hired Jenness Richardson as museum preparator. Richardson, who frequently worked in partnership with his wife, Olive Richardson, was schooled in Carl Akeley's new taxidermy methods. Akeley, the preeminent preparator of his day, had created the first "modern" diorama—an ingenious look into the lives of muskrats—at the Milwaukee Public Museum in 1887. He then moved on to work at the Field Museum in Chicago and the American Museum of Natural History in New York. Richardson went on to develop some of his own taxidermy techniques, and Olive became an expert on the methods of producing lifelike plants in wax.

The Beaver diorama depicts Minnesota's Itasca State Park. Beaver had been effectively trapped out of Minnesota during the fur trade era, and they were reintroduced to Itasca from Canada in 1901. By 1917, when work on the Beaver diorama began, Itasca's population had become firmly reestablished. Richardson spent a month in the park, preparing beaver specimens, making plaster molds of plants, and collecting enough beaver-cut wood to build a dam and lodge. Charles Corwin, one of the first artists to specialize in diorama painting, spent two weeks on site making sketches and preliminary paintings of the beaver dam and pond.

The diorama, completed in 1919, shows a family of beaver—adults nearly full-grown, yearlings, and young of the year—busy feeding and working on their dam. The background painting is not only beautiful but also clearly shows the ecological impacts of the beaver pond. Around the outside corner of the diorama, a small window allows visitors to peer into the interior of the beaver lodge and view a mother and baby beaver.

The Beaver diorama shows family members of different ages actively cutting trees, feeding on bark, and swimming in the pond.

Charles Corwin made several preparatory paintings at Itasca State Park for the Beaver diorama background mural, 1917.

With the success of the Beaver diorama, Roberts turned to another of his favorite places, Heron Lake in southwestern Minnesota. This was an amazingly productive shallow lake filled with reeds and large numbers of nesting birds and migrating waterfowl. In the late nineteenth century, market hunters shot hundreds of thousands of canvasbacks and other waterfowl at the lake and sent them off by rail to restaurants in the East. To stifle the carnage, wealthy sport hunters, including James Ford Bell, bought up property around north Heron Lake. In a progressive move, they set up hunting rules in 1901, which became the forerunners of the state's hunting regulations.

Roberts had studied breeding birds at Heron Lake and taken some of his early bird photographs there. He thought that highlighting Heron Lake would be a great way to promote conservation. At the American Museum of Natural History, his friend Frank Chapman had created the Pelican Island diorama, which helped to win support for the first national wildlife refuge at that site in 1903.

To demonstrate the amazing productivity of Heron Lake, the birds were shown in June during the height of nesting, with the adult birds incubating eggs or feeding their young. Although many of these birds nest in dense colonies, the exhibit was not completely naturalistic. It was more diagrammatic, showing all the species that would be found in a prairie marsh "community."

A local artist, H. W. Rubens, painted the background landscape of the Heron Lake diorama. In October 1920, Louis Agassiz Fuertes, the leading bird artist of the day, came to the museum and painted in the birds, sixty in all. Chapman came from New York to review this new style of diorama. Completing the foreground was a big challenge. Nests and baby birds needed to be collected and carefully prepared. Several of the birds were shown in flight about to land at their nests. The Richardsons worked for several years on bird mounts and plant models. They had to figure out a way to simulate water. Many of the nests floated in water, and many of the birds were to be shown swimming. The Richardsons concocted a thick mixture of gelatin, glycerin, dark brown coloring, and formaldehyde as a preservative. Once everything was in place, this "witch's brew" was heated and then poured onto the diorama's base, to spread and fill in around the birds, nests, and emergent plants. The diorama opened December 15, 1921. It contained sixty-five bird mounts, representing twenty-two species.

The Pipestone Prairie diorama was intended as a companion piece to Heron Lake. It showed the community of plants and animals native to upland prairies. The Pipestone location was chosen for its historic significance. For centuries Native Americans had quarried this sacred red stone and traded it throughout North America. The artist-explorer

The Heron Lake diorama included twenty-two species of birds.

George Catlin had visited the site in 1835. When diorama artist Bruce Horsfall arrived at the museum to paint the diorama background, he used Catlin's painting as a reference to depict the site as it would have appeared before European settlement. During 1925, Richardson was working on the birds and plants for this diorama as well as finishing the Bears diorama. But his relationship with Roberts had become strained. Richardson liked to keep his techniques secret and often would not let even Roberts, his boss, into his lab. Roberts found this behavior intolerable and thought Richardson should publish his techniques to spread knowledge of how to build dioramas. Richardson was also demanding more credit for his work. Roberts had allowed the background artists to sign their work, and Richardson wanted to do the same for the foreground. Roberts did not like the idea but eventually relented and let Richardson sign a rock in the foreground of the bear group. But their relationship reached a breaking point when Richardson leaked information and photos of the unfinished Pipestone diorama to the press without Roberts's approval. After consulting with other museum directors, Roberts forced Richardson to resign in March 1926.

Roberts was left in the lurch, but luckily he had a network of colleagues to draw on. It didn't take him long to find a replacement. The replacement, Walter Breckenridge, would complete the Pipestone Prairie diorama and go on to make his mark on the museum and on wildlife conservation in Minnesota.

T. S. Roberts views the Heron Lake diorama, completed in 1921.

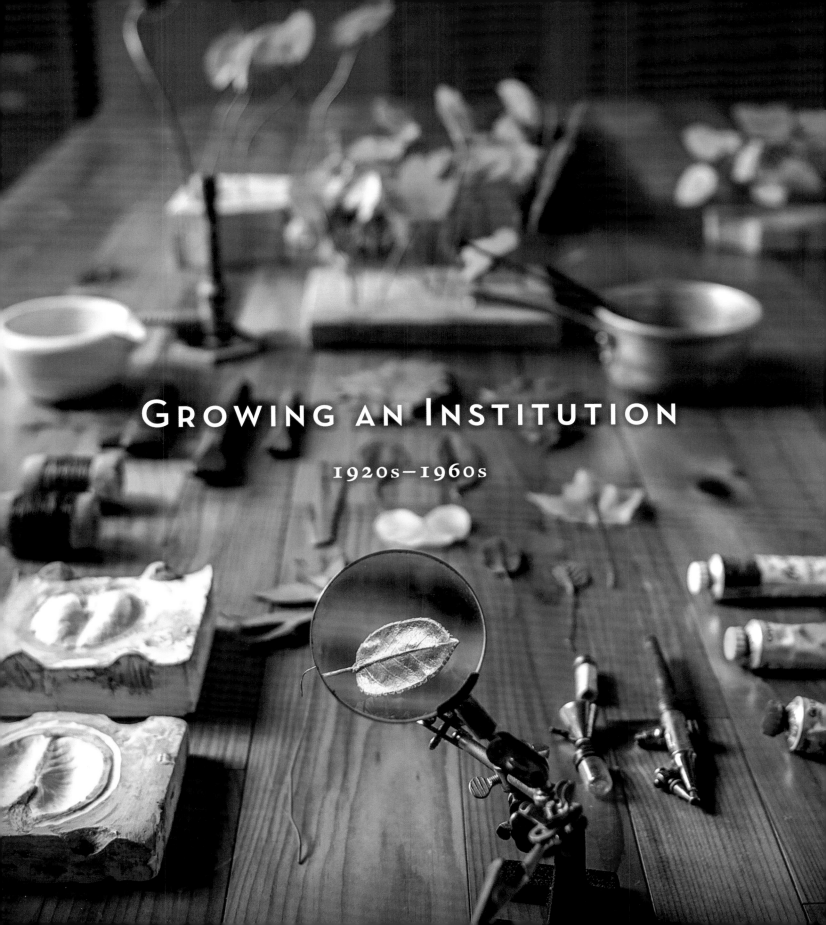

Growing an Institution

1920s–1960s

The Many Talents of Walter Breckenridge

On a pleasant April morning in 1926, Walter J. Breckenridge walked into the office of Thomas Sadler Roberts to interview for the job of museum preparator. Whatever the newly minted graduate of the University of Iowa might have been expecting, odds are he was totally unprepared for what happened next. Instead of discussing Breckenridge's qualifications or the challenges of the job, Roberts whisked him off on an early morning birding tour of Minneapolis's chain of lakes. By the time they got back, Roberts was impressed—in fact, he was so taken by the young candidate that he offered him the job on the spot without bothering to collect some essential information. Roberts later had to write to Breckenridge to get his full name, date of birth, and college major.

It was a shrewd choice, for Breckenridge—a man of multiple talents, driving energy, and dedication to advancing natural history knowledge and education—would not only succeed Roberts as museum director but would utterly transform the museum.

Growing up in the small town of Brooklyn, Iowa, "Breck," as he was known, developed an early and intense interest in nature despite a lack of opportunity. "When I was a kid, none of my teachers ever mentioned anything about the out-of-doors. Reading, writing and arithmetic were all they thought about," he recalled. "I got acquainted with outdoor things by playing hooky from school, and going down to play along the creek."

In high school, Breckenridge took a correspondence course in taxidermy, and by the time he graduated had built up a collection of sixty birds and small mammals. He donated them all to his high school, and his fellow students contributed funds to buy a display case. Breckenridge took his passion for natural history with him to the University of Iowa, where he enrolled in the nation's first college program in museum methods. He soon became the star student of Professor Homer Dill, one of the leaders of diorama creation. When Roberts wrote to Dill looking for a new preparator, Dill had no reservations about recommending Breck for the job.

Upon arriving in Minnesota, Breckenridge's first assignment was to complete the Pipestone Prairie diorama. Bruce Horsfall had finished the background, but the foreground had been left incomplete by the original preparator, Jenness Richardson, who had left the museum. Breck was faced with no small challenge. He needed to fabricate 100,000 blades of newly sprouting grass for this early spring scene. He did so by devising a machine operated with a handle and a foot pedal that enabled him to cut a tapered grass blade with midrib in one action. With this machine, Breckenridge could cut 2,500 blades in a single day. Pipestone Prairie was the last large diorama to be built in the museum's space in the Zoology Building. After completion of the project, Breck went to work on a series of small- and medium-sized dioramas and the growing collection of portable dioramas that circulated among schools.

Breckenridge was captain of the gymnastics team as an undergraduate at the University of Iowa.

Breckenridge was committed to making sure that students in Minnesota got an education that included the study of the

Detail of Pipestone Prairie diorama showing the grass blades made by Breckenridge.

natural world. Besides directing the education programs at the museum, he would go on to initiate the interpretive programs at Minnesota's state parks. Breck laid out the self-guiding nature trails and designed them to display each park's geological and biological features. Gradually the parks took over the program, but Breckenridge remained the park system's unofficial chief naturalist into the 1960s.

Soon after Breckenridge joined the museum staff, Roberts recognized that his abilities went beyond those of a preparator. The two men were soon doing field studies together, and Breckenridge took over Roberts's nature photography and filmmaking program. Breckenridge recognized that films were a dramatic new method of documenting animal behavior and educating the public. The hand-cranked cameras were heavy and unwieldy, but he was able to capture valuable footage. He made some of the first films of cranes nesting, ruffed grouse drumming, prairie chickens booming, and sharp-tailed

A sandhill crane sitting on nest in the early 1930s. Photograph by Walter Breckenridge. University of Minnesota Archives.

grouse strutting. He recalled one nightmarish attempt to film sandhill cranes. A farmer had assured him that the cranes regularly appeared each morning and evening. Breckenridge

Breckenridge with binoculars in 1935.

Breckenridge works on the Double-crested Cormorant diorama, 1940s.

spent three full days sitting beneath a corn shock awaiting their arrival. When at last the cranes landed, the camera jammed, and he sat helpless while the flock paraded within feet of his blind.

Despite these early setbacks, as cameras improved and Breckenridge's skills evolved, he became an expert at producing nature films. At first the films were shown in the museum for the popular Sunday afternoon programs. As he gained experience and the technology improved, his shows became more polished. His cameras seemed to be with him everywhere he went, including on his many trips to the Arctic. His films were among the first nature programs aired on Minnesota public television.

While fulfilling his museum duties, Breckenridge also pursued graduate degrees. For his master's thesis, he researched marsh hawks, now known as northern harriers. At the time, all hawks were considered "chicken hawks" and shot whenever possible. From 1930 to 1933, Breckenridge observed the nesting hawks from blinds and recorded their behavior and feeding habitats. He found that mammals made up 75 percent of the food taken and only one of the three pairs studied took domestic poultry.

Even though Breckenridge now had a reputation as an ornithologist, he decided to switch to herpetology, the study of reptiles and amphibians (informally known as "herps"), for his PhD research. As he traveled the state for his work on dioramas and nature films, he would come upon dead herps hit by cars on the road. He collected and preserved the specimens, noting their location. They may have been roadkills, but they greatly expanded the data on species distributions.

Breckenridge then did a detailed study of the prairie skink, one of only three lizard species found in Minnesota. In 1944, his PhD thesis was published as *Reptiles and Amphibians of Minnesota* by the University of Minnesota Press. For decades it was a popular handbook for naturalists in the Midwest. In 2014, an updated version of the book, *Amphibians and Reptiles in Minnesota*, was published as a full-color guide by noted Minnesota herpetologists John Moriarty and Carol Hall.

After Roberts died in 1946, Breckenridge was appointed director of the museum. Funding for scientific research had begun to increase in the wake of World War II. In 1947, Breckenridge moved to expand the museum's curatorial staff.

An image from Wood Duck Ways, *one of Breckenridge's most beloved movies.*

Breckenridge on snowshoes, tracking wolves in northern Minnesota, 1938.

Harvey Gunderson was hired as curator of mammals, and Dwain Warner, who had earned his PhD with noted ornithologist Arthur Allen at Cornell University, was appointed curator of birds. In the 1950s, John Tester joined the museum as an ecologist. Warner, Tester, and Breckenridge worked together to set up the University's Cedar Creek Natural History Area as a field research station. Frank McKinney, an expert on the behavior of waterfowl, was hired in the 1960s as curator of ethology (animal behavior), further expanding the reach of museum science.

Breckenridge had many wide-ranging interests. Of particular importance to him was the conservation of species and their habitats. This was the subtext of most of the Bell's popular Sunday afternoon lecture series, especially during the 1950s and 1960s, but it was an issue of concern to him almost from the beginning of his association with the museum. In the late 1940s, he was approached by the National Audubon Society to be a lecturer for the society's Screen Tour Series, a wildly popular lecture series featuring eminent naturalists narrating nature films that toured cities and towns around the country. Breckenridge's talks and use of films to illustrate the Bell's Sunday programs had attracted the society's attention, and they asked him to join their touring series. For Breckenridge, it was an opportunity to educate the public about the natural world and spread the word about the importance of conserving it. Over the next twenty years, he gave nearly six hundred lectures, visiting all but two states and reaching every province in Canada, along with several islands in the West Indies. By his estimation he wound up speaking before more than one hundred thousand people.

After Breckenridge retired from the museum in 1969, he devoted himself to his art. He produced hundreds of paintings, mostly watercolors of birds and other wildlife species. Art was not a new endeavor for him; he had always liked to draw and had taken all the art classes available when he was an undergraduate at the University of Iowa. As a preparator at the museum, he had taken time off to travel to British Columbia to study watercolor technique with the famous bird illustrator Allan Brooks. Brooks was working on illustrations for T. S. Roberts's *Birds of Minnesota*, but he could not complete all the plates in time. So Roberts had turned to a number of other artists, including Breckenridge, who composed fourteen plates of shorebirds and woodpeckers for the book.

Some of Breckenridge's favorite subjects were wood ducks. At his home on the banks of the Mississippi River north of Minneapolis, Breckenridge actively studied the nesting behavior of wood ducks using the nest boxes he had installed. Eventually he could predict to the day when a brood of ducklings would make their leaping departure from the nest. He and his wife, Dorothy, would invite friends to join them for a "coming out" breakfast. So it was appropriate that his watercolor of wood ducks would be accepted by the prestigious *Birds and Art* exhibition at the Leigh Yawkey Woodson Art Museum.

Walter J. Breckenridge, a naturalist, taxidermist, diorama maker, filmmaker, scientist, educator, conservationist, and artist, was indeed a man of many talents. After a forty-four-year career at the museum, he had almost thirty years of active retirement. His remarkable contributions were recognized in several important ways. In 1975, he received the Arthur A. Allen Award from Cornell University, one of the most prestigious honors for contributions to the field of ornithology. He was so surprised when he got the phone call that he thought it was a prank, but this was only one of many awards Breckenridge received for his numerous contributions.

And then there was the final honor. When the Bell Museum started to raise money to create an endowed professorship at the University, the decision was made to name it after Breck. In 1993, the Breckenridge Chair of Ornithology was created—a fitting tribute to a man who had done so much not only for the Bell Museum but also for the natural world around him. In 2003, Breck died at age 101.

American Kestrel, one of many watercolors of birds that Breckenridge painted throughout his long career.

Early Public Education

Reaching "the whole people..."

In 1919, when Thomas Sadler Roberts took the helm of what was then called the University of Minnesota's Zoological Museum, he arrived with big plans. Central in his thinking was the museum's role in educating the public about their natural world. While collections and displays were essential, for Roberts they were not enough. "The museum," he said, "should reach the whole people through lectures, loan collections, correspondence, and in all other possible ways."

With his friend, collaborator, colleague, and successor, Walter Breckenridge, the array of public programs Roberts proceeded to launch became the template for decades to come. Indeed, their legacy can be seen today in many of the Bell's ongoing public programs.

Roberts was a natural-born teacher. He had long enjoyed lecturing anyone who was interested in his twin passions of natural history and, of course, birds. As far back as the early years of his medical practice, he would often take time off to give talks before small audiences. Beginning in 1898, Roberts had begun embellishing his lectures with photographs he had taken, projecting them onto a screen using a "magic lantern"—a primitive slide projector whose interior light was supplied by a small generator. When he closed his practice and came to the museum, Roberts brought with him some two thousand negatives, mostly photographs of birds he had taken over the years.

His illustrated lectures became an important part of the museum's public outreach work. In the third year of his tenure, Roberts's annual report noted that he had given forty-seven lectures on and beyond campus. He would go on to deliver hundreds more over the next ten years, speaking before civic groups, public and private school classes, birding societies, Boy

Thomas Sadler Roberts photographs a chipping sparrow nest in an oak tree, 1900.

A catbird on its nest, 1905. Photograph by T. S. Roberts. University of Minnesota Archives.

T. S. Roberts and William Kilgore lead a tour of schoolchildren at the Beaver diorama, 1920s. University of Minnesota Archives.

and Girl Scouts, church groups, Y groups, women's clubs, and conservation and hunting groups. Producing these illustrated talks was no mean feat. Although the field camera he used was considered the latest in photographic technology, the equipment was heavy and cumbersome. The camera, a Long Focus Premo, was a beast made of solid mahogany. It required a tripod to be carried along on all expeditions. Captured images were transposed onto thick glass plates, which offered more permanence and better resolution than film. These, too, had to be carted around. Once developed, the black-and-white glass images were then hand-colored.

In Roberts's second year as director, James Ford Bell, the museum's chief benefactor, provided funds for a Powers movie projector. This was very early in the evolution of the nature film. Both Roberts and Bell had long recognized the potential of moving pictures to capture the lives and habits of animals in the wild, and both had been experimenting with the new medium. Moving images were captured using a heavy, hand-cranked 35 mm device. Over the next years, Roberts and his principal preparator, Jenness Richardson, generated thirty-five hundred feet of film. The films quickly found their way into Roberts's lectures and were an immediate hit, demonstrated by the increasing attendance at his public lectures. In 1920, Roberts could claim that his natural history lectures had reached more than thirty-six hundred children and adults that year.

The museum's venture into film and photography put it in the vanguard of a growing trend among natural history museums. By 1918, the museum was loaning footage to the American Museum of Natural History and the

National Association of Audubon Societies, and even the U.S. Department of Agriculture. In return these institutions loaned their films to the Bell.

One of the audiences Roberts was particularly keen to reach was schoolchildren. He was convinced that if children could be taught to appreciate the natural world as he did, when they became adults, they would be moved to protect and conserve it. "The education of the children of today. . . . is the most important part of our work," he declared. From the outset, Roberts aggressively courted area school administrators and teachers and sought to establish an ongoing relationship with them. Soon he was traveling to classrooms in the Twin Cities area, even as school group attendance at the museum rose. Schoolchildren visiting the museum were initially given a personal tour of the exhibits by the director; later, tours were led by his assistants. By then the museum was circulating films and slide packages to schools throughout Minnesota.

In 1921, Roberts began a regular special Sunday afternoon lecture series held annually through the winter months. Guest speakers came from various departments and colleges within the University and from state agencies. The talks were frequently accompanied by slides and movies, often Roberts's own footage, and were lively presentations intended to boost visitor numbers and raise the museum's public profile.

Roberts was acutely aware of new innovations and programs in other large urban natural history museums. In the early 1920s, Chicago's Field Museum had begun putting together small dioramas built for distribution to schools. This program suited Roberts's aims—in his words to "bring the museum to the public." In 1922, he sent his preparator, Jenness Richardson, to Chicago to learn how these cases were assembled and displayed, and soon he was building them at the Bell. After Richardson left the museum, Breckenridge took over the program. By 1940 the museum had 143 of these small dioramas that were available for circulating throughout the state, chiefly to schools, and during the summer to state parks.

In addition to all this, Roberts became a one-man answering service. Every day he tried to set aside time to respond, in writing, to a growing volume of correspondence from citizens wanting answers to a wide range of natural history questions. What is this bird I saw? How do I stop woodpeckers from making holes on the siding of my house?

Always alert to new possibilities to reach the public, Roberts went on the air. The University sponsored one of the nation's few student-run radio stations, WLB. Eventually Roberts hosted his own regular program on natural history, birds, and the museum. These broadcasts were sufficiently popular that he also distributed copies of his presentations to listeners who asked.

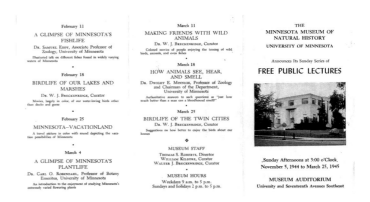

Sunday afternoon program brochure. Lectures were illustrated with slides or movies.

A portable diorama showing a snowy owl. More than one hundred of these dioramas were made and circulated to schools and libraries. University of Minnesota Archives.

Walter Breckenridge speaks to a full house in the museum auditorium, 1960s.

A state park naturalist speaks to visitors outside the Forest Inn at Itasca State Park around 1960.

After Breckenridge came on board in 1926, Roberts became less involved in public programs. He was sixty-eight and needed to spend more of his time putting together his big book, *The Birds of Minnesota*. Breckenridge rapidly became the public face of the museum, leading tours, giving lectures, and taking over the museum's public outreach functions. One of Breckenridge's initiatives was to build on the popular tradition, started by Roberts, of the Sunday afternoon lectures.

In 1936, the museum upgraded its motion picture equipment, ditching the old 35 mm camera for a new 16 mm Eastman Cine Special. Movies were now shot in full color, and Breckenridge enthusiastically embraced the new technology. He started to focus on producing motion pictures. After the museum moved into its new building on Church Street, the principal venue for his slides and movies became the new building's four-hundred-seat auditorium. As his skills improved, his films became longer and more sophisticated. Breckenridge eventually began producing seventy-minute feature-length pictures.

School group attendance at the Bell surged in the 1940s. Roberts, with Breckenridge's concurrence, expanded their small staff. Don Lewis, a biology teacher from Red Wing, Minnesota, was hired to handle bookings, lead tours, and train docents. His ability to engage the public led to his title of "public relations person," even as he became a right-hand man to Breckenridge.

With additional staff, Breckenridge found new ways to increase public outreach on natural history. In 1947, Breckenridge, now director of the museum after Roberts's death, launched Minnesota's state park interpretive program. He assigned Lewis and mammal curator Harvey Gunderson to assist the parks in designing and laying out self-guided nature trails. Starting at Itasca State Park, the program expanded to include Gooseberry Falls, Lake Shetek, and Whitewater State Parks, reaching over fifteen thousand visitors a year by 1955. Breckenridge laid out several of these trails. Gradually the Minnesota Department of Conservation's Division of Parks took over the program.

The state parks naturalist program would have a reciprocal impact on the museum. Among the state park naturalists was high school biology teacher Richard Barthelemy. Breckenridge was impressed by the man's enthusiasm and obvious love of natural history as well as his effective communication with the public. When "Bart," as he was called, returned to the University for a master's degree in science education, immersing himself in the developing field of child psychology and elementary education, Breckenridge sensed an opportunity. Barthelemy was especially interested in the emerging evidence that children learn best when given hands-on experience with things they can touch, feel, and manipulate. Breckenridge invited Bart to try out some of his ideas at the museum.

Richard Barthelemy with a school group at the Moose diorama, 1960s.

A man of unbounded imagination, Barthelemy jumped at the chance. He put together what we would call today a "learning kit"—a cart full of bones, furs, feathers, skins, and other natural objects that he had collected. Children were invited to gather round to talk about and handle the objects. This interactive approach was a great success and was, in fact, the forerunner of what nearly a decade later would become Bart's lasting legacy at the Bell: the groundbreaking museum experience known as the Touch and See Room.

After earning his master's degree, Barthelemy left the University, spending the next several years working for a variety of educational firms, where he applied his ideas about integrating the psychology of learning into the design of educational experiences. He did consulting work for the Boston Children's Museum, which was exploring new and innovative ways to engage and educate young people.

But Breckenridge never forgot him, and in 1966 he convinced Barthelemy to come back to the museum full-time in the position of public education coordinator. Bart was encouraged to see what he could do to more effectively engage the public. Possessed of an impressive imagination and steeped in the ways people learn, Barthelemy set about testing and analyzing visitor experiences at the museum. The initiatives based on these data that Barthelemy instituted over the next six years included the creation of the innovative Touch and See Room, refining the student-run, inquiry-based museum tour guide program, and devising innovative ways to draw visitors to interact with the habitat dioramas.

The arrival of Barthelemy marked a watershed moment for the Bell. What had long been coming together in bits and pieces—the movies, the public lectures, the tour guide program, and the touring small dioramas—had become a unified institutional function of the museum: a clearly defined public education program. In the decades to come, advances in technology and the readiness to embrace new means of engaging the public would enable museum visitors to explore the natural world in ways that Roberts and Breckenridge could never have dreamed. One would like to think they would be well pleased.

James Ford Bell

The Man behind the Name

On a frigid morning on January 12, 1940, with a harsh wind blowing down Central Park West, James Ford Bell mounted the great stone steps of the American Museum of Natural History. Bell was in New York City for a high-stakes meeting with the museum's director, Roy Chapman Andrews. How it would go, he hadn't a clue.

But then again, high-stakes negotiations were nothing new to Bell. At sixty, Jim Bell was a financial powerhouse as president of General Mills Corporation, a world-sprawling company he had formed in 1928 by consolidating various regional milling firms. Under his leadership, General Mills had become a hotbed of research and innovation. General Mills engineers had been put to work developing equipment that transformed simple grains into an astonishing array of new cereal products. Out of Bell's test kitchens came breakfast foods such as Wheaties and Kix. The company's Betty Crocker brand had become by 1940 an internationally recognized household name.

The man Bell had come to New York to deal with was no less formidable. Overseeing one of the world's premier natural history institutions, with vast collections and research programs stretching around the globe, Roy Chapman Andrews was no dry, retiring academic who spent his days in the museum's collection rooms. Andrews was a flamboyant and renowned explorer and fossil hunter. From 1921 to 1930, he led a series of expeditions into the Gobi Desert of China and Mongolia. Using trucks followed by supply caravans of camels and Mongolian guards on horseback, the expeditions braved sandstorms and roving gangs of bandits in a search for evidence of early humans. Andrews and his colleagues never found the early humans, but they did discover the first dinosaur eggs, the oldest mammal skull, and fossils of the largest mammal to have lived on land. After the expeditions, Andrews had become famous for his lectures, articles, and books. (Today he lives on as one of the purported inspirations for Hollywood's "Indiana Jones.")

What Bell was seeking that day in 1940 was the loan of the American Museum of Natural History's premier diorama artist, Francis Lee Jaques. Bell wanted Jaques to paint the backgrounds for a series of dioramas that were to be the central feature of the University of Minnesota's newly built natural history museum—a building for which Bell had supplied half the financing. Jaques, who was from Minnesota, was one the most renowned background artists of the day. Jaques had contacted the museum's director, Thomas Sadler Roberts, expressing his interest in returning to Minnesota to paint the new backgrounds, but he needed permission to take a leave. It was a touchy situation for Jaques. He was not on the best of terms with his boss and was under pressure to complete dioramas for the New York museum's Whitney Hall of Pacific Bird Life. He was also scheduled to join another expedition to the South Seas to collect reference materials for these dioramas. Roberts had been writing to the

In 1915, James Ford Bell was promoted to vice president of Washburn-Crosby Milling Company. He worked with farmers to increase production during the food shortages of World War I. Courtesy of Ford Bell.

American Museum for much of 1939 and had even asked his old friend Frank Chapman, ornithologist there, to intercede in securing Jaques's release. But that December Andrews had made his decision. He had written to Roberts formally rejecting his request. Bell decided it was time for an intervention.

To understand how things had gotten to this point and how important the Jaques loan was to Bell requires a bit of backstory. The diorama project was not just some rich man's vanity project. It was a critical component of a long-shared vision that Bell and Roberts had been struggling to realize for years: the completion of a modern, first-class natural history museum that would properly serve the citizens of Minnesota and the upper Midwest. Interestingly, the three principals involved in this affair—Andrews, Bell, and Roberts—all shared a common passion for natural history, exploration, and hunting, and each possessed an intense desire to convey that appreciation and concern for the natural world to the general public.

For James Ford Bell, that passion had been instilled at an early age. He was not allowed to keep a gun as a boy, so he and a friend borrowed one from his house and took their pony cart to a nearby lake, crawled out through the tall grass, and shot a wood duck. They then realized they had no way to retrieve it. Next time they took a small boat with them. Bell's father, seeing his son's interest, made arrangements for hunting trips with experienced sportsmen so he could learn the ropes. Throughout his life, at every opportunity Bell would take the train west to prairie lakes and marshes to hunt canvasbacks and other waterfowl. Over the years, as he witnessed duck populations decline, he became committed to conservation. He was also fascinated with the history of exploration and frequently visited museums to study the latest methods of natural history displays.

Unlike Andrews, who had worked his way through college doing taxidermy and had begun his lifelong career at the American Museum of Natural History by sweeping floors, Bell grew up among the wealthy milling families of Minneapolis. In 1888, his father, James Stroud Bell, had been recruited to take charge of the then floundering Washburn-Crosby Milling Company. Both the company and the family

Roy Chapman Andrews on expedition in Mongolia, 1925. He later became director of the American Museum of Natural History. Courtesy of American Museum of Natural History.

thrived under the senior Bell's direction. In 1901, at age twenty-one, James Ford Bell graduated from the University of Minnesota with a degree in chemistry. From there, he went to work for his father's company, learning the milling business from the ground up, and quickly became a force within the industry. Using his chemistry training, he set up the milling industry's first flour-testing laboratory, establishing for the first time objective standards for flour quality.

Francis Lee Jaques paints the background for the Congo bird diorama at the American Museum of Natural History, 1932. Courtesy of American Museum of Natural History.

Roberts was the Bells' family physician, and even though the doctor was twenty years older than the younger Bell, the two had been friends since the early 1900s. In the winter of 1914, when the health of Bell's aged father was failing, the family asked Roberts to join them on their yacht in Florida. For two and half months they sailed the coastlines. Roberts tended to the ailing father but still had time to fish and bird-watch. Roberts and the younger Bell tried out a new movie camera, which they used to capture Bonaparte's gulls and laughing gulls on film. When Roberts retired from his medical practice in 1915 to join the University's natural history museum, and when he took its helm in 1919, Jim Bell was right there, providing financial support and encouraging his wealthy friends to throw their support toward the museum project.

Both Roberts and Bell had become particularly fascinated with dioramas, or "habitat groups" as they were then called. This was a new way to represent nature in museums, where animals were placed into their ecological setting as opposed to standing alone in glass cases. Bell was eager to bring this type of exhibit to Minnesota, and in 1909 he led an expedition to Newfoundland to collect caribou for the museum's first diorama. It was completed in 1911. In 1914, he made an expedition to Alaska to secure Dall sheep for another diorama.

In the early 1930s, the museum was still housed in the University's Zoology Building. Bell offered to sponsor the creation of a wolf diorama. At the time, wolves had been relentlessly hunted, trapped, and poisoned to the point of extinction throughout the lower forty-eight states. Minnesota was the only place in the United States outside Alaska where wolves still survived. But to the chagrin of both Roberts and Bell, public attitudes still supported elimination of the animals, and the state paid bounties for dead wolves. When Bell presented the idea of a wolf diorama, Roberts was enthusiastic but had to inform him that there was no room for a new large diorama. Bell then offered to help finance the building of a new museum.

Bell with two Dall sheep rams taken on an Alaskan hunting expedition in 1914. The Dall sheep would become a part of a Bell Museum diorama.

T. S. Roberts did much of his early work on bird photography at Heron Lake, where James Ford Bell owned hunting property.

The Wolves diorama, completed in 1940, was the first of nine large dioramas that Francis Lee Jaques painted for the Bell Museum.

Roberts and Bell had always wanted the museum to have its own building, a destination for the public as well as a center for University teaching and research. They wanted a place where they could cultivate in the public a broad appreciation for nature and the need for its conservation. So in 1931, Bell made a formal offer to fund half the cost for building a new museum, if the state would match his gift. But the nation was in the throes of the Great Depression, and the state never appropriated the matching funds.

By 1938, Roberts, then eighty, had given up on ever seeing a new museum. But Bell, now a member of the University's Board of Regents, had not. Despite being a lifelong Republican, he thought he could get matching money from Democratic president Franklin Delano Roosevelt's Depression-era public works programs. That April, Bell invited Roberts to a meeting at his downtown office. As Roberts started to give Bell an update on events at the museum, Bell teasingly asked, "How about that wolf group?" Before Roberts could again explain to him that there was not room, Bell filled him in on the new plan. Bell had gotten the University's buy-in and had been working with the University to submit a proposal to the federal Public Works Administration for a stand-alone museum building.

By July, funding had been secured, and the architects started work in August. By December 1939, the building was complete, and the collections and exhibits were beginning to be moved in. The new building had about twice as much space for dioramas as the Zoology Building, and the galleries had been specially laid out to accommodate them. There was only one piece missing: Roberts could not get the artist he wanted to paint the backgrounds for those dioramas. That is when James Ford Bell decided to go to New York.

No one is exactly sure what transpired at the January meeting, but we do know that Roberts's friend Frank Chapman joined Bell and Andrews. Possibly at Chapman's suggestion, they decided to send a different artist on the South Seas expedition to make sketches that Jaques could later use. It was a lucky decision for Jaques; the expedition was later shipwrecked in the Coral Sea, and the crew briefly detained by the Japanese. For Minnesota, this lucky break meant that Jaques went on to paint a series of magnificent dioramas that helped make the University's natural history museum, now known as the Bell Museum, the outstanding regional museum of which Bell and Roberts had long dreamed.

In 1950, James Ford Bell formally retired from General Mills, though most of the day-to-day operations had been delegated to others during the 1940s. In 1951, the University honored him with the Distinguished Achievement Award. For the next several years, he continued to be engaged not only with the museum but with other major initiatives he had begun, including turning over his library of rare books to the University and involving himself with the world-famous Delta Waterfowl Research Station he had established on Lake Manitoba in Canada. In March 1959, Bell suffered a stroke after coming home from quail hunting at his retreat in Georgia. He died soon after. In 1967, in honor of Bell's many contributions to the museum and University, the name of the museum was changed from the Minnesota Museum of Natural History to the James Ford Bell Museum of Natural History.

As for that critical meeting in January 1940? This is all that Bell left in his diary of that day:

> 1940, Jan. 19 am—Saw Frank Chapman [and] Andrews at American Museum of Natural History. Secured their consent to release Jaques to paint some backgrounds for our museum.

Construction of a new building for the museum on the corner of University Avenue and Church Street, 1939. Folwell Hall can be seen in the background. University of Minnesota Archives.

Ogden Pleissner, Madreland, 1940s. This watercolor painting shows James Ford Bell hunting quail on his Georgia estate in the late 1940s. His grandson Ford Bell remembers riding in this wagon as a boy of eight or ten, as his grandfather pointed out all the trees, shrubs, and other flora they passed.

Heyday of the Dioramas

Windows into Nature

With the opening of the Akeley Hall of African Mammals at the American Museum of Natural History in New York City in 1936, dioramas had reached a new apogee. In this great hall, a herd of elephants confronts visitors surrounded by two floors of diorama windows. It is both a cathedral to nature and a panoramic tour of a continent.

The 1940 Bell Museum building was also designed specifically for dioramas, but it was not going to exhibit exotic locales from around the world. Instead it would focus on the diverse habitats and wildlife of Minnesota as they existed before the ax and plow. The dioramas would become a yardstick by which to measure human impact on nature close to home.

In preparation for the new museum, James Ford Bell had visited museums across the country, and with Bell Museum director Thomas Sadler Roberts's advice, he had selected Francis Lee Jaques to paint the background of the new dioramas. Jaques had established himself at the American Museum of Natural History as one of the leading diorama artists in the world, especially for birds.

In April 1940, Jaques arrived in Minnesota to work on two dioramas: wolves on the north shore of Lake Superior and migrating shorebirds on Lake Pepin. His partner in this effort was Walter Breckenridge. Since joining the museum as taxidermist and museum preparator, Breckenridge had gone on to earn a master's degree and PhD at the University. He had crisscrossed the state studying and filming wildlife. By 1940, he was probably the leading expert on Minnesota natural history. Breck, as he was known, and Jaques made a great team. Together they planned each diorama and traveled to each diorama site to research its setting and collect needed materials.

Breckenridge would handle the taxidermy and other foreground preparation, such as making molds of rocks, collecting plants and trees, and making wax flowers and foliage. The overall design was mostly left up to Jaques. They created exhibits that not only placed animals into natural settings but also told stories about how animals live in nature. In the dioramas of many natural history museums, big-game mammals were typically posed before a backdrop, as if in a formal family portrait. Instead, Jaques and Breckenridge strove to create windows into a dynamic natural world where animals and plants interacted in a natural system. A family of wolves is clearly hunting, and if viewers look closely, they will find the intended prey, a deer, painted into the background. The moose is not standing erect looking out at the viewer but is

Akeley Hall at the American Museum of Natural History in New York set the standard for museums across the country: an open central hall surrounded by dioramas. Courtesy of American Museum of Natural History.

Francis Lee Jaques spent a few months in Minnesota in the spring of 1940. He visited the North Shore and Lake Pepin to make sketches and take a few photographs, and then painted both background diorama murals before returning to his job in New York.

walking into the scene, following and hoping to mate with the female in the background. In each diorama are a story to discover and something about to happen.

For Jaques, working on these dioramas was a great homecoming. In his fifteen years at the American Museum, he had traveled the world re-creating scenes from nature to bring back to New York. From coral reefs in the Bahamas and tropical rain forests of Panama, to windswept tundra and mountain vastness, he had sought to capture the essence of each environment. But now he was coming home to the places he knew and loved. One can sense that feeling in his work.

One of the skills that set Jaques apart from other artists was his ability to paint birds in flight. He gave them space in which to move—vast expanses of sky that always told a story of approaching weather. He was a master at creating the illusion of depth by modulating contrast and the intensity of colors. He also used tricks to draw the viewer into the scene and into the story, such as a path into the woods, or the water lines along a beach. Jaques was not interested in creating a photographic reproduction of a scene. He had a distinctive artistic style in which he removed all unnecessary detail and distilled animals

Walter Breckenridge puts the finishing touches on the wolf mounts, 1940.

and landscapes into essential shapes and forms. These elements were assembled like pieces of an ecological puzzle. A Jaques diorama is not just a wonderful window into nature, it becomes an icon of an ecosystem. The great bird artist Roger Tory Peterson, who met Jaques soon after he came to New York, recognized his achievement: "Jaques brought the diorama to the highest degree of development."

Snow Geese diorama, completed in 1942.

Sandhill Cranes diorama, completed in 1946.

Many people consider the Snow Geese diorama the finest bird diorama in the country. It has several unique innovations. The raking sunlight is achieved by placing a series of lamps along the left side of the window. Jaques carries that same sidelighting onto the birds in the sky, beautifully re-creating the evening light as it plays across the perfectly structured plumage. The patterns and groupings of the birds are masterfully depicted, with the positions of the wings varied in such a way as to create a sense of motion. The water, a chromium-plated copper sheet, buckles slightly to create a wavy reflection. As the viewer moves, the reflection changes, suggesting the motion of water.

Like in the Snow Geese diorama, the sandhill cranes are depicted during their spectacular migration. Both greater and lesser subspecies of crane are shown. In the 1940s, cranes were rarer than they are today, so Breckenridge revived and restored some from old study skins rather than collecting new specimens. The cranes are shown in native prairie habitat, bowing, leaping, and picking up sticks, all parts of their courtship display. While visiting this diorama's site in the Red River valley,

Jaques taped paper cutouts to the wall when designing the placement of the cranes for the background mural of the Sandhill Cranes diorama.

Jaques and Breck noticed a three-rut track in the grass. They believed this to be a remnant of the Pembina (Red River) oxcart trail that traders used to carry bison hides from Winnipeg to St. Paul in the 1830s. They decided to include it in the corner of the background. Today, this site is protected as a Minnesota

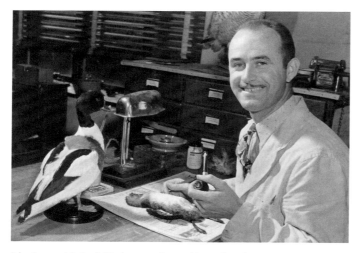

John Jarosz with the shelduck mount that got him a job at the museum.

Museum staff in front of the Big Woods (Maple–Basswood Forest) diorama, 1949. Though Jarosz did the taxidermy and most of the foreground preparation work, both Ruth Self (left) and Dorothy Mierow (right) worked with him to make hundreds of botanically accurate wax plants.

Scientific and Natural Area. The rest of the Red River valley is almost entirely converted to agriculture or other human activity.

During the height of World War II, the Bell Museum received a mounted shelduck, a European duck species, which was confiscated by customs for lack of proper permits. When Breckenridge went to pick up the duck, he recognized it as some of the best taxidermy he had ever seen. He traced the origin of the mount to John Jarosz, a local boy from northeast Minneapolis, who did taxidermy as a hobby. Jarosz was stationed in England when he shot the bird on a hunting outing. In England, he mounted the bird using his penknife and scrounged materials. He then mailed it to his parents. Based on this one piece of work, Breckenridge wrote to Jarosz offering him a job as soon as he returned from the war. Jarosz proved to be a natural as a museum preparator. His mounted animals were not only exquisitely prepared but also beautifully captured the animals' actions and expressions. After Breckenridge was appointed museum director in 1946, most diorama foregrounds were prepared by Jarosz.

Jarosz worked with Jaques on completing the last four large dioramas: Elk, Tundra Swans, Big Woods (Maple–Basswood Forest), and Cascade River (Coniferous Forest). The Cascade River was the last large diorama to be created and is one of the museum's most beloved. It does not have an obvious animal subject and may appear at first glance to be empty, but closer examination reveals an amazing diversity of wildlife. By making many of the animals inconspicuous, this diorama stimulates observational skills and the sensation of discovery. Many small birds are sounding alarm calls in the branches surrounding a great gray owl hiding in the spruce tree. This mobbing behavior is a common reaction to owls and other potential predators. In such encounters, for a few brief moments, the surrounding foliage becomes filled with noisy birds before they disperse and disappear.

The visitor may be startled to spot a newborn fawn curled up at the tree's base, a reaction not unlike that of coming upon a real fawn in the woods. The fawn is a good example of Jarosz's skillful taxidermy. Accidentally killed by a farmer cutting hay, the fawn is missing both its front legs, but Jarosz arranged it in such a position that the missing legs are concealed. The upper

These wax models of young basswood leaves were made for the Big Woods (Maple–Basswood Forest) diorama. It took two years to complete all the wax leaves and wildflowers for this diorama.

Jaques applies black underpainting to the Cascade River (Coniferous Forest) diorama.

The Cascade River (Coniferous Forest) diorama was painted in 1954 and completed in 1956.

A bull moose strides through a muddy shore on Gunflint Lake in a diorama completed in 1946.

falls of the Cascade River form the central spectacle of the background painting. The falls are made more dramatic by the deep shadow on the near cliff. To achieve this high contrast, Jaques underpainted the dark area with flat black paint. High contrast and fine details are used selectively to tie the background to the foreground on the left side. Through the artistry of Jaques and the craftsmanship of Jarosz, visitors are invited to witness the intimate workings of nature.

The 1940s and 1950s were the heyday for dioramas at Minnesota's natural history museum. In total, Jaques in partnership with Jarosz and Breckenridge completed nine large dioramas and ten smaller dioramas (see appendix, "The Bell Museum Dioramas"). The dioramas depict Minnesota's native biomes, from native prairie to deciduous woodlands to coniferous forests and rocky shorelines. Within these habitats, visitors can witness some of Minnesota's iconic species, such as the moose, wolf, and loon, as well as explore Minnesota's amazing diversity of neotropical migratory birds and the now extinct passenger pigeon. The dioramas were state-of-the-art museum exhibits in the mid-1900s, and they remain exquisite works of art, created by some of the best diorama artists in the field.

THE DIORAMA DILEMMA

Although dioramas were icons of natural history museums for many decades, by the 1970s their value was being questioned from several directions. By design, wildlife dioramas almost always included little or no indication of human presence. They were presented as views of nature before the

axe and plow of European settlers and the impact of modern development. Empty of humans, they also wrote Native Americans out of the picture—implying that Indigenous people were not present or had little to do with the land. When natural history museums did include Native people in dioramas, they appeared to be equated with wildlife.

Dioramas were also seen as an impediment to expanded science teaching. Educators wondered just how much visitors were learning when they looked at a diorama. Was building an affinity for nature enough, or did museums need to deliver more hard-core science content? Static, frozen-in-time dioramas began facing increasing competition from other experiences. People could watch wildlife nature programs at home on television and later on the internet and other venues. The rapid expansion of science centers, often equipped with IMAX theaters and dynamic, interactive exhibits, further deepened these challenges.

The Bell Museum, with its cadre of student interpretive guides, continued to create engaging and popular experiences for school groups at the dioramas, using them to discuss contemporary environmental and cultural issues. But for many visitors, simply looking at the windows and reading labels were no longer sufficient to hold their attention or provoke their curiosity. Though the museum experimented over the years with ways to augment the diorama halls, a true reinvention of the diorama experience was needed. But that would have to wait until the dioramas were moved to a new building in 2018.

Taking Flight

The Artistic Journey of Francis Lee Jaques

In 1924, Bell Museum director Thomas Sadler Roberts received a letter from a commercial illustrator in Duluth seeking a job as a museum artist. It was a fine letter, but the writer had made a huge faux pas. He suggested that he could improve some of Roberts's early bird photographs by touching up the blurry wing tips. Retouching photographs for scientific publication, however, was considered fakery. Roberts had no trouble writing him a discouraging reply.

The letter was from Francis Lee Jaques, an artist who went on to become the foremost diorama painter of his time and whose stunning background scenes for the Bell became a signature feature of the museum. But in 1924 he was unknown, and his letter to Roberts was his first attempt to leave his hardscrabble life in northern Minnesota. Jaques had grown up on failing farms in Kansas and had moved as a teenager with his family to a log cabin north of Aitkin, Minnesota. He had worked as a lumberjack, taxidermist, railroad fireman, and electrician before being drafted into the military during World War I. He had a passion for nature, particularly birds, and an innate talent to draw. Throughout his childhood he drew and painted, and as an adult he taught himself to be an artist with help from Clarence Rosenkranz, a professional artist in Duluth.

Nursing his disappointment over the rejection from Roberts, Jaques went duck hunting. While eating lunch, he startled a black duck, which flew up from the grasses next to his boat. Inspired by the beauty of the bird's flight and its motion through the reeds, he made a painting that not only depicted the bird accurately but also captured the essence of the moment and the passion Jaques felt for the bird's marsh environment. Deciding he had nothing to lose, he sent the painting to Frank Chapman at the American Museum of Natural History in New York. Chapman was a leading ornithologist and conservationist working at one of the world's greatest museums. Chapman was amazed by the painting and invited Jaques to join the museum's staff.

At age thirty-seven Jaques moved to New York to start a storied career painting diorama backgrounds. At the time he understood bird flight better than any other artist, and he knew how to integrate birds into broad landscapes. He was soon joining scientists on museum expeditions to Panama, South America, the Bahamas, Alaska, the Arctic, and the South Pacific. He also began illustrating books on birds and natural history. He worked with some of the leading scientists of the day, including Ernst Mayr, who would go on to apply Darwinian theory to the developing field of genetics to form today's understanding of evolution.

In New York, Jaques met and married the aspiring writer Florence Page. For a honeymoon, Jaques took Florence to Minnesota for a canoe trip in his

Francis Lee Jaques was drafted in 1918 and shipped to San Francisco for artillery training. He visited the California Academy of Natural Sciences, where new dioramas painted by Charles Corwin had recently been completed. Jaques was awestruck, and this experience planted the seed for his future career.

Francis Lee Jaques, Just before Dawn—Pintails, *oil on canvas, date unknown. Memories of duck hunting on early mornings with his father left a vivid impression on Jaques.*

Francis Lee Jaques, Black Duck Stamp Design, *watercolor wash, 1940. Jaques was asked to create the federal Duck Stamp in 1940. His composition is often considered one of the program's most elegant designs.*

In 1934, Jaques traveled aboard a sailing schooner to many remote islands during the Zaca expedition. Here he makes reference paintings for a diorama on the Galapagos Islands. Courtesy of California Academy of Sciences.

Jaques on the Zaca expedition in the South Pacific, 1934. Courtesy of California Academy of Sciences.

Lee and Florence Jaques at their home in Minnesota in the early 1960s. Photograph by Noel Dunn.

beloved boundary waters (now the Boundary Waters Canoe Area Wilderness). Out of this experience came the classic book *Canoe Country*, written by Florence and illustrated by Lee, one of the first popular books written about this region. Florence and Lee went on to write and illustrate six books about their travels together. When renowned nature writer Sigurd Olson started to write about the value of wilderness and the need for its preservation, he turned to Jaques to illustrate his books *The Singing Wilderness*, *Listening Point*, and *The Lonely Land*. Besides the twenty dioramas he painted for the Bell Museum, Jaques illustrated more than forty books and completed some of the most beautiful plates for T. S. Roberts's 1932 book *The Birds of Minnesota*—an ironic coda, of sorts, for a long and close relationship that had begun so inauspiciously for Jaques many years before.

Many artists are superb at painting birds or other animals. And many artists have mastered the art of painting landscapes. But Francis Lee Jaques was one of the first to combine the two, to truly integrate animals into their environments. He did this with a style uniquely his own. He had a penchant for selectivity—emphasizing the most important elements of scene and subject—which he combined with a masterful sense of design. He defined his subjects, whether trees, water, birds, or mountainsides, with a few lines or silhouettes. The finished drawing would be representational yet also a work of strong abstract design.

Each element in a Jaques painting is distilled to its essential form. These components, like interlocking pieces of a puzzle, are sensitively presented according to their aesthetic and ecological relationships. In the finished painting, the distinct parts are integrated in a composition that expresses a vision of nature as an ecological system. In Jaques's work, we see the natural environment presented as a vital but endangered heritage, painted for its own intrinsic beauty and significance. His achievement was an artistic, visual expression of the American wilderness ethic.

Florence Jaques had never been in a canoe or even gone camping before their honeymoon trip to the boundary waters (now the Boundary Waters Canoe Area Wilderness). She later developed her journal into the book Canoe Country.

Francis Lee Jaques's painting of passenger pigeons and mourning doves for The Birds of Minnesota *demonstrates his exquisite sense of design.*

Francis Lee Jaques, Caribou on Ice, oil on canvas, 1940s. As a mist of ice crystals fills the air, woodland caribou walk along the frozen shore of Gunflint Lake. Jaques painted caribou as a missing ecological piece of the northern forest ecosystem.

Wildlife Explorations

1940s–1980s

At the Poles

Arctic and Antarctic Research

Under the guiding hand of Thomas Sadler Roberts, the University's natural history museum had become nationally known and a highly regarded, state-focused museum. But after Walter Breckenridge became the museum's director, it became something more. Under his leadership museum scientists began moving out into the larger world, exploring the ecological connections between the flora and fauna of the upper Midwest and those of subtropical Central America and the earth's polar regions.

TO THE ARCTIC

From practically the beginning of his employment at the museum as a preparator, Breckenridge seemed to be drawn to the Arctic. He wound up making several collecting and filming forays into some of the most remote parts of these landscapes.

In 1953, Breckenridge got his first opportunity to organize a museum-led research expedition to the high Arctic. It was to the Back River in Canada's Northwest Territories. The expedition was sponsored by Robert and James Wilkie, who owned a machine business in Savage, Minnesota. Also in the party were University geologist Philip Taylor, museum preparator John Jarosz, and the museum's mammalogist Harvey Gunderson. They were to explore a section of the Back River that had been visited only twice before by outsiders—first by British explorer George Back in 1834 and then by James Anderson in 1855.

Breck and his party arrived by floatplanes on July 12, just as the ice on the lakes was breaking up. They were hoping to witness the large herds of caribou and musk oxen that had been reported by both Back and Anderson, but they found none. The party collected birds, small mammals, plants, and geological specimens to add to the museum's holdings. They also had a grant from the U.S. Army to test several new insect repellents. Of course, Breck recorded all these activities on film.

John Jarosz and Walter Breckenridge prepare bird specimens during the Back River expedition, 1953.

Walter Breckenridge and the Wilkie brothers film and make sound recordings of a Hudsonian curlew at its nest, 1953.

During their stay, the party encountered several groups of Inuit, who followed a traditional way of life based on fishing in summer and caribou hunting in winter. Forty years later, Breck got a request from Canada to review his films. One of the Inuit whom Breck had met and filmed on the 1953 trip had become a renowned artist, and Breck's films were rare images of the artist's early life.

The trip was not without its dangers. During the trip's last week, the floatplane arrived early to take out much of the expedition's collected specimens and some equipment. It was to return a few days later to pick up the party and the rest of their equipment. The Wilkie brothers, who were eager to return home, decided they would take advantage of the plane's presence to fly out with the gear. The University men still had work to do and chose to remain—a fortuitous decision for them, as it turned out.

On the way back, the plane developed engine trouble. The pilot guided the plane down to a lake, but it hit a rock reef, collapsing one of the pontoons, and the plane began to sink. The men managed to get out and somehow stabilized the craft. Fashioning a raft out of parts of the plane, they made it to shore with their sleeping bags but with no means to communicate to the outside world. They were there for three days when, using an aluminum specimen tray as a mirror, they were able to signal a distant bush plane. Within hours they were picked up. Later, a U.S. Army Air Force plane returned to the Back River and flew out the rest of the party.

Breck continued to be drawn to the wildlife of the Arctic. He made two trips to Wales, Alaska, on the Bering Strait, in 1964 and 1965. On the second trip, Breckenridge and outdoor writer Jim Kimball had an opportunity to join an Inuit party on their way back to Little Diomede Island. The rocky surface of Little Diomede Island is famous for its colonies of nesting sea birds, and Breck jumped at the chance. The only catch was that the Inuit hunters had no idea of when they might be coming back to Wales. Breck and Jim packed their cameras, other equipment, and about a week's supply of food, and climbed into an umiak, an open boat built from driftwood and walrus skins.

As the Inuit maneuvered the skin boat through fog and mist, walruses were spotted on the ice floes, and the trip became a hunt. The hunters used .22-caliber rifles to bring down these four-thousand-pound animals. A kill required a perfect shot to the ear. Soon hunters were jumping on and off ice floes and desperately trying to harpoon the dead animals before they sank. At one point, an angry bull walrus charged the boat. The journey turned out to be a bigger adventure than the island itself. A week later, on the trip back to the mainland, Breck was glad when no walruses were spotted!

Jarosz and an Inuit family outside their caribou-skin tent on the Back River, 1953.

Back River expedition preparing to depart Inuit village on July 28, 1953.

Breckenridge filming during a storm near Wales, Alaska, 1964.

THE BELL GOES SOUTH

By the end of the 1960s, and as part of the growing worldwide concern for the planet's environment and its resident species, the nature and direction of biological research at the Bell Museum expanded. Museum researchers turned their attention to the southern polar regions.

In 1968, at the urging of George Llano, director of the National Science Foundation's Antarctic biology program, the Bell began an ongoing study of seal populations along the coast of Antarctica. Llano had learned of the museum's early work in radio-tagging wildlife and wanted to begin using this method to study the wildlife of Antarctica. Under the direction of the museum's curator of mammals, Al Erickson, and newly hired statistician Don Siniff, museum researchers supported by the U.S.'s McMurdo Station began counting and tracking Weddell seals off Ross Island. Developing and refining the technology as they went, they later expanded the research to include other seal species as well.

But perhaps the person who did the most to advance the museum's polar research during these early years was ornithologist David Parmelee. Born in Oshkosh, Wisconsin, Parmelee grew up in Iron Mountain, Michigan. After a stint in the Marine Corps in the South Pacific during World War II, he went on to earn his undergraduate degree at Lawrence University in Wisconsin, a masters at the University of Michigan, and his PhD in ornithology from the University of Oklahoma, where he was a star student of legendary ornithologist and bird artist George Miksch Sutton, who introduced him to the Arctic. Over the next twenty years, Parmelee became a world-class expert on the birds and ecology of the Arctic region.

In 1970, the University of Minnesota hired Parmelee to direct its prestigious field biology programs at the University's Lake Itasca Forestry and Biological Station and the Cedar Creek Natural History Area. In taking the job, he was obliged to abandon any further studies in the Arctic, since the season for field research and teaching in Minnesota conflicted with the summer field season in the Arctic. Determined to continue his polar work, he turned his attention to Antarctica, where the summer field season coincided with winter in Minnesota, when fieldwork was shut down. Two years after coming to Minnesota, Parmelee was spending the winter months in Antarctica and very soon became a leading expert in the birds of that region as well.

The bulk of Parmelee's Antarctic research centered around the U.S. station on the Palmer Archipelago, a chain

David Parmelee holds a one-year-old wandering albatross that is still too young to fly.

of ice-bound islands between the Antarctic Peninsula and the Weddell Sea. Parmelee considered the program he set up there to be the most important of his accomplishments, for it gave his students an opportunity not only to hone their research skills but also to interact with students and researchers from all over the world. Over the course of his twenty years in the Arctic and another twenty in Antarctica, Parmelee mentored hundreds of students, many of whom went on to become noted researchers and teachers in their own right.

"He always let us know he was proud of us and our accomplishments," remembered one of his grad students, Pamela Pietz. "He consulted us and referred others to us as though we were tenured colleagues. This not only built our confidence, it also deepened our sense of responsibility and our desire to do our best."

Parmelee was in many ways a kind of scientist from an earlier era—equal parts gentleman naturalist, artist, and explorer—but one whose work also met the highest scientific standard of the time. His prepared specimens were models of perfection. He was a renowned collector of bird nests and eggs and was known for an uncanny ability to find nests where others could not. He authored more than 120 professional and popular scientific articles, two monographs, and two books. An accomplished photographer and artist like his mentor, Sutton, Parmelee produced works in watercolor, pen and ink, and pencil that are beautiful in both their detail and composition. A trove of some of his finest works is published in his last and most impressive book, *Antarctic Birds: Ecological and Behavioral Approaches*.

While he was director of field biology, he was responsible for dramatically upgrading and improving the physical and scientific facilities at Itasca and Cedar Creek. Parmelee also greatly expanded and improved the University's Biology Colloquium, a program designed to introduce freshman and sophomores to biology outside the classroom. In the words of one veteran ecologist who had gone through the program and had become hooked on the science, "The Colloquium had a profound influence on many budding biologists."

When Dwain Warner, the Bell's curator of birds, retired in 1986, Parmelee was offered the job, which he eagerly accepted. Under his leadership the museum's entire bird collection was brought up to date and computerized.

Although he was decidedly "old school" in his approach to research, demanding of himself (and his students) long hours of close observation and attention to detail, he was also open to new, innovative technologies. In 1985, he collaborated in the first attempts to track seabirds by satellite.

In honor of his work and accomplishments in Antarctica, a mountain was named for him: the Parmelee Massif, a massive wall of sedimentary and metamorphic rock and ice along the east coast of Palmer Land.

David Parmelee's watercolor of chinstrap penguins, a good example of his talent as an ornithological artist. Courtesy of Helen Gale Bruzzone (Parmelee).

The Bride Wore... Boots?

Nothing quite says, "I love you" like a pair of hiking boots. Well, that is, if you are Walter Breckenridge. In keeping with the consummate field biologist and collector that he was, Breckenridge's approach to marriage and a honeymoon was in total agreement with his professional passions.

Breckenridge met Dorothy Shogren, his future wife, at a blind date picnic on Memorial Day in 1933. Six weeks later they were engaged and planning to be married later that year, after Breckenridge returned from a seven-week collecting trip to the Canadian Arctic. He had been invited to help University of Minnesota bacteriologists Robert and Beryl Green collect rabbits, grouse, and other wildlife that were carriers of ticks known to transmit tularemia, a human disease. The Greens were interested in determining the northern limits of the disease.

The day before they were to leave, Breckenridge introduced the Greens to his fiancée. Surprised that Breck had a girlfriend, let alone a fiancée, the Greens said that had they known, they would have invited her to come along, too. At a farewell dinner that evening with Dorothy's mother and sister, Elizabeth, and another couple (a minister and his wife), Dorothy mentioned their earlier conversation with the Greens. The minister's wife suggested a plan: why didn't the two of them get married *before* Breck left, then use the trip as their honeymoon? Her husband, she suggested, could do the officiating. It was a ludicrous idea, of course. Breckenridge was leaving in less than twenty-four hours. But by the end of evening they decided, Why not?

Walter and Dorothy Breckenridge on their wedding day in 1933. They left on an Arctic expedition the same day.

The next day featured a month's worth of planning and preparation crammed into less than twelve hours. First, Breckenridge called the Greens to verify that they meant what they had said about Dorothy coming along. He then had to get permission from his boss, Thomas Sadler Roberts. A friend of Dorothy's offered her place as a wedding venue.

At seven o'clock that morning Dorothy went to work as usual at the First National Bank in St. Paul, where she was to await the call from Breck that the plan was a "go." Meanwhile, her mother and sister packed her things in a duffel bag—including rice, confetti, and a new pair of hiking boots wrapped with pretty white ribbon (a gift Breck had given her the week before). When Dorothy got the call that afternoon, she walked into her boss's office, quit her job, and then went shopping for long underwear, flannel pajamas, and an air mattress. She met Breckenridge a short time later, and the two bought a ring, after which they drove to the friend's house, where the minister's wife had arranged for flowers and refreshments. The wedding took place at four thirty that afternoon, T. S. Roberts and Dorothy's sister, Elizabeth, serving as attendants. An hour later, the newlyweds were on a train bound for Winnipeg to begin their "honeymoon."

It was, as they say, a trip to remember, one for which Dorothy's home economics degree had done little to prepare her. Three days later, this former manager of the First National Bank's Crossways Tea Room found herself out running traplines and shooting grouse along the shores of Cormorant Lake, where they stopped en route to Churchill, Manitoba, some four hundred rail miles to the north.

Dorothy Breckenridge carrying water at their research camp in northern Manitoba. University of Minnesota Archives.

Walter Breckenridge works at the research camp. University of Minnesota Archives.

Dorothy Breckenridge with moose antlers found near their camp. University of Minnesota Archives.

Breckenridge prepares specimens on the deck of a boat on Hudson Bay, 1933.

On August 14, they arrived in Churchill, where the group's real test was to begin. A few days after they arrived and set up their tents, a prolonged stretch of severe weather struck. Driving rain and high winds soon forced them to move their work base and living arrangements to more secure quarters—a tar-paper shack at the edge of town. Over the next few weeks, the two couples ranged up and down the lower Churchill River, collecting ptarmigan, snowshoe hares, lemmings, mice, and other wildlife. At one point, Breckenridge left the group. Chartering a small yacht captained by a local man and his son, Breckenridge and a man hired as a cook motored up the coast of Hudson Bay nearly a hundred miles and then canoed to a small Inuit encampment, where Breckenridge was able to collect a few more specimens of mice, birds, and plants. The way back turned into a real struggle as high winds and rain drove a violent chop that tossed the little party around for much of the trip before they landed safely, much to the relief of Dorothy and the Greens.

Weather continued to be a challenge, but they all did get through it. In the end, no scientific report was ever published regarding the northern limits of tularemia. But Breckenridge got what he wanted, a host of specimens for the growing museum collections. And Dorothy got a good taste of what life was going to be like for the next sixty-nine years as the wife of Walter Breckenridge!

Migrations

The Life and Times of Dwain Warner

As the Bell began to expand the scope of its research into the wider world, the museum was fortunate to find scientists with the talent and vision to take it there. One of those individuals was Dwain Willard Warner. Born in 1917, Warner grew up on a dairy farm in Northfield, Minnesota, and attended Carleton College there. He received an undergraduate degree in botany and zoology before going on to Cornell University to study ornithology. At Cornell, he signed on to a select bird-collecting expedition to the Tamaulipas region of northeastern Mexico led by two icons of the professional birding world, George M. Sutton and Olin S. Pettingill Jr. An accomplished wing shooter, Warner became an indispensable member of the group. The experience fired his interest in the birds of Mexico, which became a lifelong passion and a major focus of his research.

But the year was 1941, when the United States entered World War II. Warner left his studies and enlisted in the army.

Dwain Warner holds a boat-billed heron while George Sutton makes a sketch during their expedition to the Tamaulipas region of northeastern Mexico, 1941. Courtesy of Richard Warner.

He was quickly shipped off to the South Pacific, as the army had taken note of his background in the biological sciences and the fact that he was a crack shot. Stationed in New Caledonia, the New Hebrides, and New Zealand, he was assigned to an elite special biological team, ostensibly to work on rat research and malaria control. In fact, he and his team spent much of their time hunting wild game for the troops stationed in the area. But Warner also took advantage of his free time to bag and preserve the local birds whenever he got the chance, adding them to his growing collection. After returning to Cornell in 1946 following the war, Warner switched his dissertation research from the birds of the Mexican state of Tamaulipas to the birds of New Caledonia and the Loyalty Islands, basing much of his thesis on his wartime field experiences.

In 1947, Warner returned to Minnesota, joining what was then the Minnesota Museum of Natural History (forerunner of the Bell) as an associate professor in zoology and the museum's first official curator of ornithology. At the museum, he was able to return to his original passion, the birds of Mexico, although teaching demands required his presence in Minnesota during the regular school year, leaving little time for research. But he made the most of the summers, heading to Mexico for intensive fieldwork and collecting. Every few years, he would take his entire family with him. Providing his children with shotguns and nets, he enlisted them in his collecting work.

An important contact for Warner in Mexico was the world-renowned mammologist Bernardo Villa Ramírez, a legendary figure in Mexican science. The two became close friends. In 1972, in collaboration with Ramírez's home institution, the National University of Mexico, Warner set up a biological research station in the Los Tuxtlas region in the southern state of Veracruz. Here, in a topographically and

Biological research team in New Caledonia during World War II. Dwain Warner kneels in the foreground. Courtesy of Richard Warner.

ecologically complex landscape, lay the most northern fringe of the Southern Hemisphere's tropical rain forest, still in a relatively intact state. Over the next twenty years, the station became the principal hub for the study of the region's rain forest birds, expanding what was known about their distribution, behavior, and migration ecology and the effects of changing land use and habitat loss.

During the latter half of his career, Warner was able to extend his research to the region's nonmigrating bird populations. When he was not in Mexico or holding classes on the Bell Museum's top floor, Warner was teaching field ornithology at the University's Lake Itasca Forestry and Biological Station.

Warner was known for his highly creative approach to research—his penchant for thinking "outside the box." He enthusiastically embraced new technologies that became

Warner with bird collection in New Caledonia, 1944. Courtesy of Richard Warner.

available after the war and sought ways to apply them to his science. He was an early advocate of the use of transistors to remotely monitor wildlife. Teaming up with engineers, he was one of the first to experiment with radio telemetry. In 1962, he applied (unsuccessfully) to the National Aeronautics and Space Administration for funds to develop a satellite-based wildlife tracking system. In this he was ahead of his time by thirty years.

Warner was one of those teachers whom everyone wishes they had had. Extremely personable, he had an energy and enthusiasm for his work that was infectious. Possessed of a deliberate irreverence that frequently left his students laughing, he made learning highly entertaining. He was also demanding. "Field trips," recalled one former student, "would typically begin at dawn in a bog somewhere and end at dark watching woodcock display from a position as close as the birds would allow." Along the way, his students would learn not only about the birds they were hearing and seeing but also about the biology, geology, and often the human history of the landscape through which they traveled. Whether watching Warner stand on the roof of a van under a sky full of migrating waterfowl and offer a beer to the first student to spot a white-fronted goose, or hearing him on a bird-collecting expedition in a far-off land exhort his grad students to "pump 'em full of sunshine," accompanying Warner in the field could be a life-changing experience.

The man had a voracious need to know not only about the lives and habits of the birds he studied but about the wider world in which they occurred. And he enthusiastically shared every new discovery. This extended to his own family. "Everywhere we'd go, he was always looking at things, pointing things out about what we were seeing and experiencing," recalled his son Richard. "From vegetation changes we were seeing as we gained altitude, to the rock strata we saw in road cuts, and even to the languages and dialects spoken in towns we went through. Every trip was a lesson about the world around us."

Warner was a true innovator on many fronts. His research into the behavior and distribution of neotropical bird residents of southern Mexico provided a valuable foundation for later workers, as did his research on migration ecology. He and his graduate students did much to advance Mexican ornithology and were among the first to document habitat destruction on the wintering grounds of many North American breeding birds and its effects on their populations.

Never one to remain idle, after his retirement in 1987 Warner used his professional connections to lead more than a dozen safaris to Kenya. Over the course of his forty years at the museum, he published more than one hundred papers and oversaw the expansion of the museum's ornithological collection by nearly twenty-eight thousand specimens, turning it into one of national significance. He died in 2005 at the age of eighty-eight.

Warner in his office at the Bell Museum, 1980s.

In the 1960s, Warner collaborated with Allen Downs, a University of Minnesota professor of studio arts and a well-known photographer and filmmaker. They shared a passion for Mexico and worked together to document the natural habitats of the Veracruz region. Photograph by Allen Downs.

Tracking Nature

The Rise of Wildlife Telemetry

Little known today is that for one shining decade, from the mid-1970s to the mid-1980s, the Bell Museum was the world's leading developer of radio-tracking and monitoring equipment for tracking wild animals in their natural habitats—the forerunner to today's practice of remote sensing by satellite. Housing a team of design engineers and biologists, the museum's lab at the University's Cedar Creek Natural History Area was the international go-to center for researchers seeking to track everything from Antarctic seals, to salmon, to African lions, to polar bears, wild horses, rats, ducks, tigers, and a host of other species.

The idea was hatched in 1958 when the Soviet Union's Sputnik satellite program launched Laika, a dog, into space. Laika was wired to a transmitter that enabled scientists on the ground to keep track of her vital signs during the brief few hours she was alive. It was, in effect, the first demonstration of radio telemetry's capabilities, and the experiment lit a fire under the museum's curator of ornithology, Dwain Warner.

Warner had spent a lot of time in Mexico studying the ecology and habitat use of several species of Central American birds that migrate north in the spring to breed in Minnesota and beyond. He recognized immediately the usefulness the technology could have not only to his own work but also to the work of researchers everywhere who had long sought ways to penetrate the daily lives of wild animals in the natural world. If the technology could be applied to dogs in space, surely it could be applicable to rabbits or birds here on Earth. With the help of friends, Warner contacted the University of Minnesota's dean of the Institute of Technology, Athelstan Spilhaus, to see if he might help. Spilhaus was intrigued and agreed to support Warner in the search for funds.

But developing the technology was only half the challenge. Warner needed a research site to test the technology—a place where animals were living free in their natural environment. He decided to approach Arthur Wilcox, director of the University's Cedar Creek Natural History Area, a 5,400-acre site thirty-five miles north of the Twin Cities, to seek permission to use the property to build out the system and test its use.

Warner's timing was opportune as the University was already in the midst of building a field station on the site. Warner's idea was to install environmental sensors in various habitats on the site. These would be connected by a network of buried electrical cables to the laboratory in the new building. Wilcox was convinced. A joint proposal was sent to the National Science

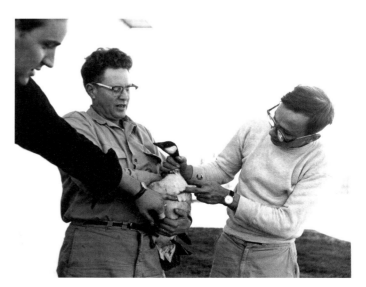

Dwain Warner, Bill Cochran, and a student attach a radio tag to a Canada Goose, 1962.

Cottontail rabbit with radio transmitter at Cedar Creek Lab, 1963.

Foundation, which came through that same year with a grant of $55,800 to cover the cost of installing cables and environmental sensors. The project was off and running.

This was all outside-the-box stuff. The application of space technology to ecological research had the air of the bizarre about it, and finding research dollars to advance the science was not easy. There was no assurance that the technology could be successfully applied to animals in the wild, let alone yield information that was in any way better than what could be gained using old-fashioned field biology techniques. By this time, wildlife biologists in other states had begun exploring telemetry's use for game management. Papers were beginning to published in respected professional journals, and there were now special sessions set aside at conferences for those interested in using the new technology. There was even a *Wildlife Telemetry Newsletter* published by the Wildlife Society.

One of those states was Illinois, where biologists working with the Illinois Natural History Survey, in collaboration with William Cochran, an engineer with experience in transistors, had developed transmitters small enough to fit a rabbit. Rabbits had been released and successfully tracked. The results were reported in the Wildlife Society's telemetry newsletter, and the word quickly spread. Warner and Bell Museum researcher John Tester had come out with their own prototype and were impressed by Cochran's simple lightweight tag, which was far superior to their own. In fall 1961, they hired him. Cochran's arrival put the Bell Museum on the inside track to develop the technology.

Cochran was a workaholic and brilliantly creative. His first project upon arriving in Minnesota was designing and building a handheld receiver and a set of transmitters that were fitted to three captive white-tailed deer. The animals were then released and tracked over a ten-day period. With that success under his belt, Cochran told Warner and Tester he could automate the system using the facilities at Cedar Creek. Given the go-ahead, Cochran hired Larry Kuechle, an electrical engineering graduate student from the University's Institute of Technology, to help him put the system together. Kuechle was a perfect hire. A detail man with a knack for solving technical problems, he was instrumental in turning Cochran's designs into functioning equipment. He was particularly good at working with the biologists using the system, tweaking and refining the equipment to accommodate their research needs.

THE PROJECT TAKES OFF

In March 1963, at the North American Wildlife Conference's telemetry session, Vincent Schultz, chief ecologist of the U.S. Atomic Energy Commission (AEC), announced to the standing-room-only crowd that the AEC was open to proposals for the development of telemetry technology for wildlife research. Tester was there and saw an opportunity to get long-term financing for the automated system. He returned to Minnesota and submitted the first of what would be many successful proposals to the AEC that would keep the Cedar Creek telemetry project going for years. On December 31 that same year, Cochran's automated system was turned on, and Minnesota's telemetry project went into high gear.

The build-out of the Cedar Creek system over the next two years was, in the words of one of the many scientists who came through the program, an exercise in "organized chaos"—not surprising since the system was being used even as it was being expanded, tested and refined. Parts of the system were forever breaking down, towers were in constant need of maintenance, and receivers and transmitters needed adjustment and continual improvement.

One of two towers with directional antennas that located tagged animals by triangulation.

A bank of radio receivers at the Cedar Creek lab that tracked the movements of multiple animals, each identified by a unique frequency.

One of the greatest challenges was how to manage the tsunami of data being generated. To handle it, extra staff were hired to read the films and tabulate animal positions. It took six hours to plot the movements of a single animal over a twenty-four-hour period—three hours to transcribe the locations from the film, and three hours to plot them on a map. The data were on thousands of punch cards, which were stored at the Bell Museum and read by a machine. These would have to be carted by wheelbarrow over to the University's computer center, where someone would make sense of it all. In 1964, Don Siniff, a statistician and biologist, who would go on to become one of the University's celebrated Antarctic researchers, was hired to analyze the data. His analytical skills significantly expanded the services the project team was able to offer to researchers.

Having successfully gotten the Cedar Creek system up and running, Cochran left in 1964 to return to Illinois, and Kuechle took over as director of the project. But the groundwork had been laid, and in the following years the system and its various components continued to be refined and improved. The Cedar Creek engineers were constantly redesigning the equipment to accommodate the research needs not only of University researchers but also of visiting scientists from beyond Minnesota who were looking to the Cedar Creek group to come up with the tags and receivers that would work with the species they were studying. Mobile receiving equipment was constantly being improved in response to researchers' requests to adapt it for use from aircraft and watercraft for studies in remote areas far from Cedar Creek.

The bioelectronics lab staff found an increasing amount of their work located away from Cedar Creek. Operating as contracted consultants, they supplied equipment and planning in support of the research needs of institutions around the world, including studies of Weddell seals in Antarctica, wild horses in Nevada and Oregon, lions in Africa, salmon on the Snake River in Washington, steelhead trout in Lake Superior, and polar bears in the high Arctic.

BEGINNING OF THE END

Times and circumstances change and, with them, fortunes. Although the contracts with external researchers brought in some money, the automated operating system was the financial lifeblood of the lab—it brought in the federal grants. After the passage of major federal environmental legislation during the 1970s, particularly, the Endangered Species Act, federal and state money became more and more targeted to applied research and the solving of species-specific problems. In addition, the group found itself competing with for-profit equipment manufacturers and retailers of telemetry technology.

Alan Sargeant plots animal locations on a map of Cedar Creek Natural History Area, 1964.

Green-winged teal with transmitter.

Then in the late 1970s, the funding landscape began to shift. As privately funded tracking technology advanced, the technical limitations of the lab's automated system and its small, two-and-a-half-mile range were becoming increasingly obvious. At the same time, federal agencies, the main funders of telemetry studies, were less inclined to fund development of the technology, as concern rose that the focus on such gadgetry had grown at the expense of scientific study.

The automated operating system that had provided the bulk of the group's financial support now seemed more like an anachronism. Advanced computer-based data collection and analysis made the cumbersome film-based system increasingly obsolete. Funds to maintain and update the system became harder to find. Kuechle and his core staff became increasingly concerned about the long-term financial viability of the lab.

In 1981, Kuechle and some of his staff decided to form their own for-profit company to continue the work of equipment manufacturing. But the University, concerned that for-profit work would threaten the integrity of the research site, asked the group to take their business off the premises. After Kuechle moved the operation off-site, his staff, one by one, left to join the company. Kuechle chose to remain at Cedar Creek, taking a leave of absence from the company. But then, in 1990, he too left Cedar Creek for good. With his departure, the Cedar Creek project came to an end.

Alan Sargeant, the first biologist hired by Warner, in 1963, perhaps best summed up the significance of the telemetry project. It was not so much the system itself, he said, or even the data that it generated that were so important. "Rather, it was the impact its development and successful use had on stimulating worldwide interest in the radio tracking of animals. The subsequent widespread application of this technology," he concluded, "has been phenomenal."

Don Siniff and Larry Kuechle use a bag to capture and hold a Weddell seal in Antarctica while gluing a tracking device to its fur, 1970s. Courtesy of Larry Kuechle.

Mystery of the Missing Toads

In 1957, John Tester, a young ecology grad student working on his PhD, was trying to solve a mystery. He had been documenting the effects of burning, grazing, and mowing on the species makeup and ecology of Waubun Prairie, a tract of wet prairie in northwestern Minnesota. This was a classic, tallgrass prairie ecosystem, a landscape of high grass peppered with small wetlands and ponds fringed by bulrushes and sedges. Big bluestem and Indiangrass dominated the uplands, which were pocked here and there by brushy hummocks called "Mima mounds"—small knolls of weedy forbs and wolfberry bushes. It was a 250-acre patch of land teeming with life, none more conspicuously present than a small, noisy amphibian known as the Canadian toad.

In early spring, adult males congregate in ponds and wetlands to form breeding choruses. In the evening their raucous calling can be deafening as they attempt to attract females. For the rest of the spring and early summer, both males and females, along with the young of the year, move to pond and wetland shorelines. Then, around late August and early September, the toads fall silent. In fact, they disappear. "During that first spring and summer they seemed to be everywhere," Tester recalled. "And then I couldn't find them anywhere. One day they were everywhere, and then they weren't anywhere."

Puzzled, Tester decided to find out what had happened. His idea was to catch and radio-tag a few toads in the spring so he could follow their movements through the year. Radio tagging involved injecting small bits of radioactive tantalum wire under the toads' skin. The animals' movements could then be tracked by handheld scintillation counters—a kind of small Geiger counter. But Tester knew little about the toad's behavior, so he decided to seek the help of the state's foremost expert on amphibians, Walter Breckenridge, director of the Minnesota Museum of Natural History (now the Bell Museum).

Breckenridge needed little convincing. Under a grant from the Atomic Energy Commission, that next spring the two men tagged eight toads and began tracking their movements. But then, in late August, they lost the signal . . . and the

Above: Canadian toad, scratchboard illustration by Don Luce.

Walter Breckenridge locates marked toads with a Geiger counter.

John Tester measures the depth of hibernating toads.

The Canadian toad, a species found in the prairies of western Minnesota.

Mima mound with fences set up to capture toads. Photograph by John Tester.

toads. Firing up their counters, the two men began searching all the areas where they thought the animals should be. "We first zeroed in on the shorelines, but got no signal," said Tester. "We thought, well, they might have gone into the mud at the bottom of the ponds. But when we got out onto the water and turned on the equipment there was still nothing."

"So we decided to get more systematic about it. We started at the edge of [one of the major ponds] and started walking around it in ever-widening circles until we reached one of the Mima mounds." There the counters came alive. When the two scientists started to dig, they soon came up with toads. In the end, they managed to retrieve seven of the eight they had tagged.

Further investigations revealed the toads were using the mounds almost exclusively as their wintering sites. Some of these hummocks contained more than three thousand toads!

Over the next several years the two scientists were able to put together a more complete picture. The evidence they gathered strongly suggested that the mounds were a consequence not only of the burrowing toads but of other ecological forces as well. Pocket gophers, ground squirrels, and other burrowing animals were also moving the earth. These smaller mammals in turn attracted larger burrowing predators, such as badgers, who further churned up the soil. All this loosening and mixing of soil in turn led to a change in plant succession above ground, from grasses to forbs to shrubs, all of which led to further permeability of the soil—in effect, creating "puffs" of well-aerated soil on the flat prairie landscape. Toads, which cannot survive frost, take advantage of these places that offer ease of movement through the soil, enabling them to move deeper underground as soil temperatures fall in the winter, and higher in the soil as soil temperatures rise in the spring.

The Museum in the Environmental Era

1960s–1990s

Touch and See

Innovating Hands-On Learning

Rare is the person who can be said to single-handedly bring a new way of thinking to the work of an institution. But a case can be made for Richard Barthelemy, the Bell Museum's public education coordinator from 1966 to 1972. He turned the museum visit for kids from a passive encounter into an exciting hands-on learning experience and set the museum's public education programs on a lasting trajectory. Fifty years later, Barthelemy's legacy continues.

Walter Breckenridge was already aware of Barthelemy's talent when he hired him, having observed his innovative teaching style when he was a graduate student in the 1950s and working part-time at the museum. Now an expert in the psychology of learning and its role in the design of educational experiences, "Bart," as he was known, was given free rein by Breckenridge to do what he could to increase the museum's attendance and make the visitor experience more meaningful.

One of Barthelemy's first initiatives was to conduct a visitor survey to find out just who was coming through the museum's doors. The survey found that more than 80 percent of those visitors were children ten years old or younger. This, he decided, was a problem for the museum, as its exhibits were designed to appeal primarily to a more mature audience. He set about making the museum more kid friendly. His first programs continued an activity he had tested at the Bell while a graduate student. He spread items from the museum's collections—bones, beaks, fur, feathers, and other artifacts—on carts or directly on the carpet and invited visiting children

Richard Barthelemy with a group of schoolchildren in the Touch and See Room, 1972. He experimented with placing a "tent" over a particular specimen, in this case a turkey skeleton, to focus the group's attention.

The elephant skull was—and still is today—one of the great mystery objects in the Touch and See Room. Visitors young and old are encouraged to inspect it and to try to figure out what it is.

to study and handle them. The activity was a huge hit with kids. When a large room in the museum's new wing became available in 1968, Barthelemy saw an opportunity to take this to another level. He moved his trove into the more spacious quarters, calling it the Touch and See Room.

One of the first of its kind in the nation, the room was uniquely designed to generate inquiry, and its objects were placed and presented to foster discovery. Everything in the room—the objects, their positioning, where they were located—was carefully considered, down to the soft tan colors of the walls. Collaborating with Bart in choosing where things would go and how they were presented was University of Minnesota educational psychologist Roger Johnson and Barthelemy's artist wife, Margaret. Together, they turned the space into an intensely tactile and visual space. The floor was covered with wall-to-wall carpeting. Moveable walls enabled the manipulation of space and light. Modular tables offered the opportunity to move exhibits and work spaces around to accommodate whatever programs staff wanted to showcase. Children's movements throughout the room were closely mapped, and all exhibited objects and images were presented at a child's level.

Barthelemy was constantly interacting with kids and staff, analyzing what was working and what was not. He was

A dissecting microscope was added in the 1980s to allow magnified viewing of many of the room's objects.

A student guide shows a class of young students a snake. After the discussion, each student gets a chance to touch the snake.

Live animals were added to the room in the 1970s and are still among the features in the new Touch & See Lab. Photograph by Joe Szurszewski.

forever making adjustments and changes to the physical space and the way in which staff interacted with young visitors. As is true to this day, nothing was labeled; this was a room for discovery and questioning. He believed in open-ended, active learning that starts with questions and triggers curiosity. University students hired as interpretive guides were trained by Bart in the art of inquiry-based learning.

Bart was also a master at generating publicity, using his contacts with local news outlets to attract public interest for the museum's programs. The room quickly became one of the museum's main attractions, dramatically expanding young attendance. For school tour groups from across the state, the Bell became a favorite destination. The room also became a model for other museums throughout the country, including the Smithsonian Institution's National Museum of Natural History.

In the years that followed, the Touch and See Room experienced further changes. In 1976, it underwent a dramatic remodeling by then curator of exhibits Terry Chase. Graphic panels bearing images of classic scientific illustrations were hung on the walls. The tan-colored walls were repainted black, in effect, turning the room into a black-box theater, which made both the panels and the room's featured natural history objects stand out. As the taxidermied animals became worn through constant touching and handling, they were restored or replaced, and objects were constantly rearranged or changed.

In the 1970s, live animals were added to the room—snakes, turtles, and later a variety of invertebrates, such as hissing cockroaches, giant millipedes, and fiddler crabs. In keeping with the room's theme, all the animals could be touched by visitors, with the supervision of the student interpreters.

But the Touch and See Room was not just a favorite place for children. In the 1980s, the room and its objects were also used for public classes in natural history illustration. In fact, the room became a studio of sorts for art students from the University and other colleges in the metropolitan area. Later, a once-a-month Sketch Night program, mostly attended by adults, featured changing themes from the museum's collections.

Jennifer Menken, public collections manager, created a teaching collection in the early 2000s built from donations and surplus specimens from the Bell's scientific collections. This collection supplied the need for changing objects on display, served as a source for public education programs, and became a part of a library-like loan program for local classroom teachers. In fact, in recognition of the value and importance of the teaching collection, it was given its own space with large tables, the Collection Cove, in a room adjoining the Touch & See Lab in the new museum.

Today the Touch and See Room lives on. An updated version, the Touch & See Lab, has been incorporated into the new Bell Museum, including some of the original objects, now more than fifty years old, like the elephant skull and the deer skeleton, and new favorites, like the group-friendly digital microscopes. It remains an enchanted place, waiting to be discovered and explored by a new generation of children and by visitors of all ages who are eager to interact with real objects and curious about the amazing natural world around them.

Jennifer Menken, Touch & See Lab coordinator, enjoys an interaction with a visitor at the fur table. Menken played a critical role in redesigning the room for the new museum. Photograph by Joe Szurszewski.

The Touch & See Lab in the new Bell Museum includes upgraded technology, such as this digital microscope. Photograph by Andy Hardman.

Public Programs

From Education to Engagement

School is not where most Americans learn most of their science. . . . an ever-growing body of evidence demonstrates that most science is learned outside of school. —"The 95% Solution," American Scientist

It has been said that in nature it is "adapt or die." Much the same could be said of institutions. Take, for instance, the Bell Museum and its public programs.

Traditionally, natural history museums had focused on introducing visitors to the wonders and ways of the natural world. But in the 1970s, they faced a very different reality: mounting evidence that humans were causing serious damage to the earth's ecosystems, air, and waters. In response, a national movement arose that sparked the founding of Earth Day and major environmental legislation.

For natural history museums this was a watershed moment. For the Bell Museum in particular, this was a challenge that had to be met. Dedicated to educating the public about the natural world and nurturing an appreciation for its protection, the museum's mission gained new urgency. The Bell responded by expanding the content of its public programs to include the major environmental issues of the day.

Previously, the museum's public offerings had focused on tours for school groups and Sunday afternoon films and lectures aimed at adults and family groups. Responding to the dramatic rise of interest in nature and science fueled by the environmental movement, Bell programs grew to include a wildlife information call-in line, answers to natural history questions in the local newspaper's Mr. Fixit column, a first-rate natural history bookstore (the Blue Heron Bookshop), informal evening classes, field trips led by graduate students, and multiple programs for all ages.

When Gordon Murdock joined the staff as curator of public education in 1981, he organized public conferences on topics such as prairie conservation and climate change, brought in big-name speakers such as Jane Goodall and Jared Diamond, developed teacher training workshops, and oversaw the beginning of the museum's flagship summer camps. Over the next three decades he administered an expanding offering of public

During the 1980s, the Bell Museum helped to bring nationally recognized speakers, such as Jane Goodall, to campus. She spoke at Northrop Auditorium, and receptions were held at the Bell Museum.

Informal evening courses and field trips, often taught by museum graduate students, were popular in the 1980s and 1990s.

programs. He eventually assumed leadership of the University's museum studies graduate minor, started earlier by Bell Museum director Don Gilbertson and the heads of the University's Weisman Art Museum and Goldstein Gallery. This program connects the public functions of the University's museums to its core instructional work. To better publicize the museum's growing offerings, the quarterly newsletter IMPRINT was started in 1984, including a detailed calendar of events.

In 1990, the Bell Museum became a regional center for the JASON Project, an effort to improve science education through the use of new technology. Created by Robert Ballard, a celebrated deep-sea explorer, this two-week electronic field trip enabled school students to interact in real time with scientists working in remote locations, like the Galapagos Islands. The success of JASON encouraged the museum to develop its own version, called Bell *LIVE!*, which ran annually from 1994 to 2000.

When the scientific collections and their curators moved to the new Ecology Building on the St. Paul campus in 1993, the museum took advantage of the space they vacated to create classrooms for a wider array of public programs. Classroom space meant school groups touring the museum could also have in-depth laboratory sessions to enrich their tour experience. Summer camps, a highly popular program, also took advantage of the classroom spaces. To reach schools statewide, the museum developed the Schott Learning Kit program. Teachers could select from a wide variety of topics, such as "Exotic Aquatics" and "Amphibians and Reptiles." Each kit included a detailed curriculum, activities, reference books, and a set of authentic objects for study, all packed into a large trunk.

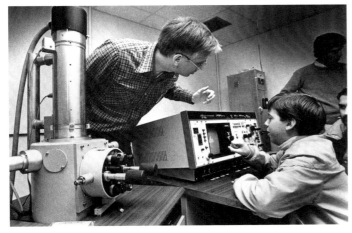

As part of the University of Minnesota, the Bell Museum was able to offer unusual programs, such as tours to the electron microscope lab on the St. Paul campus.

Gordon Murdock, the museum's curator of public education.

In the 1980s, weeklong summer camps for kids six to twelve years old were a new offering. Forty years later, more than eight hundred kids attend these widely popular programs each summer.

Kevin Williams and Shanai Matteson host a Café Scientifique at the Kitty Kat Club, a pub in Dinkytown near the museum, in 2007.

From 1990 to 2003 more than fifteen thousand students filled the Bell Museum's auditorium for two weeks every spring to watch, via satellite hookup, scientists exploring the natural world as part of the nationally produced JASON Project.

NEW CHALLENGES FORCE A NEW DIRECTION

By 2000, new challenges forced the museum to reevaluate its educational goals and mission. Local demographic shifts meant that future audiences for museums were going to be significantly more diverse. Moreover, the rise of the internet and advances in research about learning were upending long-held assumptions about how visitors used museums. With access to information at people's fingertips, the mission of museums as sources of information had new competition. At the same time, visitor research showed that museums were not particularly effective at conveying detailed factual information. Instead, studies showed that visitors came to museums for social interaction, to see real, unusual, and amazing things, and to have an experience that stimulated reverence and wonder. But the importance of museums remained undisputed. Other research indicated that visitors still *did* learn in museums and that Americans gained as much as 95 percent of their science knowledge from informal education sources, such as museums.

As part of a museum-wide effort to reinvent itself, Bell educators began to investigate how best to focus efforts and resources. How could the museum attract and communicate with new audiences? Could museum programs help to alleviate issues such as the achievement gap? Given the museum's

Museum curator Keith Barker shows a group of Girl Scouts a white-throated sparrow during a Saturday with a Scientist program. The bird was banded and released.

Starting in 2013, the Bell held an annual picnic-in-the-winter Valentine's Day event in the diorama halls.

limited assets, was there a unique niche that the museum could fill in the community?

One of the museum's greatest assets, its connection to the University, was now increasingly engaged in creative ways. The Honeybee Program, for example, highlighted the University's expertise in bee research and created a learning experience that was effective in reaching urban school children. Another popular program, *Saturday with a Scientist*, featured University scientists and attracted all ages—the young, the old, and family groups.

In 2003, the museum joined an international movement to take science communication into the community to places where people gather for recreation, including bars and cafés. Shanai Matteson, who started as a student intern, was determined to make the museum "cool" to young adults, a group who otherwise rarely visit natural history museums. Matteson organized a monthly program, Café Scientifique, where a University of Minnesota scientist spoke in a local bar to enthusiastic crowds. Later she created the *Bell Social*, an event with live music and food that combined an art exhibition by the current Bell resident artist with a presentation by a University researcher.

Over the decades, the Bell Museum has continued to offer a diverse array of educational programs for the public. Bell educators keep finding that "sweet spot" for museum

From 2004 to 2018, the Bell Museum coordinated an annual BioBlitz program. Bell Museum curators, along with other University scientists and resource managers from the Minnesota Department of Natural Resources, would identify as many species as possible from one location during a twenty-four-hour period.

programs by featuring those elements unique to the Bell, including its connection to the wealth of researchers and resources of the University of Minnesota, by integrating art with science, by offering a direct, personal experience with real objects from the natural world, by engaging visitors in the process of discovery, and by offering a relevant connection to its Minnesota home.

George Rysgaard, a student guide, gives a tour to a visiting school group at the recently completed Wolves diorama, early 1940s.

Interpreting Nature

The Student Guide Program

"I'm curious," says the young woman with the name badge on her University of Minnesota sweatshirt. "What do you think is going on here?" She is standing before one of the Bell Museum's renowned dioramas—a big bull elk bellowing across a wooded savanna below him. The little knot of people around her—an elderly man with a young girl in tow, a couple from out of state, a teenager with earbuds draped around his neck—collectively lean in toward the scene. Other questions follow: What is the elk doing? Why is the elk bugling? Can you find the red-tailed hawk? The questions and offered answers quickly turn into a conversation that opens up a world far more complex than the scene before them. One of the things that sets the Bell Museum apart from others of its kind is its time-honored interpretive guide program.

In 1940, when the University's Minnesota Museum of Natural History opened the doors to its new building on Church Street, it was one of the few places in Minnesota outside the classroom where the public could find information about Minnesota wildlife and the state's native habitats, and other important facts about the state's natural history. Central to the museum's public education program then and to the present were its magnificent dioramas. But the dioramas alone were not enough. Just as important were the interpretive staff who were there to guide visitors through those dioramas, to help visitors understand the layered realities they depicted—the ecological connections among the displayed species and the behavior of the featured animals. In other museums this role was usually played by adult volunteers, often retirees with time on their hands.

But early on, the museum had decided to do something different. Taking advantage of its position within a research and teaching university, it offered students (primarily undergraduates studying the natural sciences) part-time employment as interpretive guides. With access to talented and motivated University students, the museum had the uncommon opportunity to offer visitors a personalized, small-group experience.

The creative genius behind this kind of communicating was Richard Barthelemy, the museum's first public education coordinator. A learning innovator in the early 1960s, Barthelemy trained student guides in a new way of interacting with visitor groups. Instead of *telling* visitors what they were looking at, interpretive guides would ask questions of visitors and then would have them search the dioramas for answers. A similar technique was used with children; they were challenged to explain objects they found in the Touch and See Room. The method quickly became the museum's signature teaching style.

The interpretive program proved effective and popular, and the public responded by flocking to the Bell—especially school groups, which would regularly fill the halls of the museum's two main exhibit floors. But things began to change in the 1970s with the rise of the environmental movement, and the Bell found itself competing for an audience. The museum had to respond. What the Bell had going for it was its intimate space and the profoundly personal connection it offered visitors through its student guides. Visitors were guaranteed a customized, open-ended experience.

Student interpretive guides were asked to create their own interactive programming. One early innovation was getting younger visitors engaged by role playing. Guides would get kids to pretend they were certain animals in the dioramas—an exercise that became a lesson in animal behavior.

A child enters the front doors of the Church Street museum.

A group of children come to the museum for a guided tour, 1980s.

In the 1990s, a student guide engages a school group in active role playing—soaring like an eagle.

Jennifer Menken was a student guide in the 1990s. "We really were unique in how we embraced active role play," she recalled. "This worked really well because of how the museum was set up. Small groups could engage with the dioramas without disrupting the other groups on the floor." One of the favorites was the crane mating "dance," which kids would perform before the Sandhill Cranes diorama.

For older visitors, guides were encouraged to go beyond the basic information about the scenes before them. They were urged to draw on what they were learning in their University classes and to experiment and come up with their own innovations, a daring policy move at that time. In fact, the freedom granted to the guides was tremendously stimulating both to them and to the visitors with whom they interacted.

Most students came to the guide job with at least some knowledge about science and natural history since they were studying ecology, evolution, and wildlife biology, but the guide position attracted students from other disciplines as well. During her time as a tour guide, Jennifer Menken was enrolled in the University's fisheries and wildlife program. "The [guides] represented a wide diversity of colleges," she said. "Many came from the College of Natural Resources such as myself. . . . but we also had education, geography, anthropology, and dance majors."

Training for the guides became more structured. By the 1990s, the guides met monthly for more in-depth instruction. Curators presented updates on the latest research in their fields, and specialists from the University's education department lectured on inquiry-based learning. Guides met periodically during the school year to go over what was working and what wasn't and to share what they were learning in their studies that might be useful in the museum galleries.

A student guide with a class in front of the Moose diorama. Guides were trained to use the inquiry approach: to ask questions to stimulate thinking.

A student guide uses a worksheet with prompts and questions designed for older school groups.

But as important as the student guide program was for museum visitors, it turned out to be just as important for the student guides. Many who signed up had already decided on a career in the natural sciences, but others didn't know what future they wanted. They just knew they were interested in the museum and its subject matter. As they got to know the science and the wonders of the natural world, many became hooked and went on to pursue careers in a broad range of environmental science–related professions.

Bell guide alumni include two directors of the National Eagle Center, Mary Beth Garrigan and Rolf Thompson. Barbara Hansen, a tour guide in the 1960s, became a palynologist (expert in the study of pollen and spores), while Robin Andrews became a professor at Virginia Tech, specializing in ecology and evolution. Dave Palmquist, a tour guide from the 1970s, went on to a thirty-year career working for the Minnesota Department of Natural Resources' parks division, and Jeb Barzen became director of field research for the International Crane Foundation. Chris Goodwin, a guide in the 1990s, joined a recycling business that serves Twin City households.

Adele Binning, who was a student guide in the 1980s, could speak for many about what the tour guide stint meant for a young undergraduate. "It had such a profound influence on so many people that have walked through those doors," she said. "I could . . . list many people who have gone on to be field biologists, who have gone on to lead efforts to preserve natural areas, who have gone on to educate hundreds and hundreds of children." Binning became an exhibit developer for the Science Museum of Minnesota and, later, a science teacher.

Over the years, the guide program has grown steadily from a force of fifteen to twenty interpreters in the 1970s to more than fifty at the opening of the expanded new Bell Museum in 2018 on the St. Paul campus. The museum depends on its student staff. Their roles include leading tours, staffing the Touch & See Lab, teaching summer camps, and presenting live planetarium shows.

In addition to having made a critical contribution to the museum, former guides find the experience memorable and unifying. Some guides keep in touch long after their Bell Museum experience. "People I worked with, we still get together for a birding breakfast in the spring," said Sue Buettgen, who was a tour guide in the 1980s. "I really do think in our jobs now we carry a little bit of the Bell with us."

A student guide at the new Bell Museum with a summer camp group. Photograph by Joe Szurszewski.

From Student Guide to College Professor

When Anna Pidgeon first signed up to be a Bell Museum interpretive guide, she was a college kid with only a general idea of what she wanted to do with her life. Today she is a professor in the University of Wisconsin's Department of Forest and Wildlife Ecology. Her story is not unlike many students whose Bell interpretive guide experience helped spark their own careers. Recalls Anna: "My goal was to work as a biologist for some nongame program. Being a tour guide, I thought, was fun and a great way to use what I was learning in school while making a little money."

In 1983, Anna was an undergrad in wildlife management at the University of Minnesota when she heard about the tour guide job from a fellow student. She inquired and was hired by Kate Murray, then head of the program. "She was a very creative type in drama and community theater, and she got tour guides doing as much movement activities with children as possible. She trained me in," said Anna.

The tour guide job was one of several paid-by-the-hour gigs Anna juggled while getting her undergraduate degree. She briefly left the Bell Museum to do other interpretive work for the Hennepin County Parks Department. She also began taking evening classes in education, reasoning that what she learned could be applied to whatever wildlife management job she would eventually get. But these were the Reagan years, and wildlife management jobs were not easy to come by. When Kate Murray left the museum in 1985, Anna applied for and got Kate's job: boss of the interpreters. She remained in the position for eight years and led the development of new program initiatives that, decades later, are still in place.

Anna Pidgeon with schoolchildren in the Touch and See Room, 1980s.

"Initially I just did what I had learned to do: maintain the Touch and See Room, conduct in-service training for guides, and organize seasonal weekend programs for families," she recalled. As she gained mastery over the work, she began thinking about other ways of doing outreach.

"I was involved in the Minnesota Naturalists' Association when I worked for the Parks Department. I got a lot of ideas through the workshops that they would put on," she said. Anna began running overnight stays at the museum for kids and began a Bell-based summer camp program (over three decades later, summer camps are one of the museum's flagship programs). Together with the Bell's curator of education, Gordon Murdock, and the Minnesota Department of Natural Resource's Nongame Program, she organized conferences on Minnesota-focused topics, such as native prairies, and partnered with the Minneapolis theater company In the Heart of the Beast for a theme-based Halloween program.

But in the end, it was Anna's interactions with the scientific staff that led her to what was to become her career. Drawn to their work out of curiosity, she began hanging out with the grad students and their advisors, the Bell Museum curators. "I found rubbing elbows with the curators to be just so exciting," she said. She volunteered to work with some of the grad students on their projects—offering her labor for free, and learning all the while. Eventually, she became a kind of virtual grad student, working unfunded for the museum's curator of mammals, Elmer Birney, who became a mentor to her. "I learned a ton," she said, "and attempted to communicate that [to the public]."

But before long she began itching for more. "What you are able to interpret here, most of the time, is at a really basic level," she said. "And I'm seeing all this cool information that people are gathering . . . I said, I want to be able to make my own information, too."

In 1993 she took the plunge, entering a master's program at Central Washington University studying northern spotted owls, followed immediately by entrance to a PhD program in wildlife ecology at the University of Wisconsin, where she researched birds of the Chihuahuan Desert. Her vocation was sealed.

Today, Professor Pidgeon's research specialty is avian ecology; she teaches ornithology to hundreds of students and is mentor to her own cohort of graduate students. Her experience at the Bell Museum, she readily allows, was a profound influence in that journey. "It made me," she said, "see what was possible."

Anna Pidgeon teaches field biology techniques as a professor at the University of Wisconsin, Madison, 2019. Courtesy of Anna Pidgeon.

Making Movies
Reaching a Bigger Audience

The big-screen television at the front of the classroom briefly flickers, then suddenly fills with the piercing glare of an American bald eagle. *"Live from the Raptor Center on the University of Minnesota St. Paul Campus,"* comes a voice out of the ether. *"Bell LIVE! Productions presents Raptors LIVE!"* A collage of faces of other birds of prey appears, followed by a blizzard of fast-cut scenes of scientists at work. *"Come along on our electronic field trip and see the behind-the-scenes rehabilitation of injured eagles, hawks, falcons, and owls. Watch diagnoses, treatment, and physical therapy as veterinarians and biologists care for these injured birds of prey . . . ask our scientists and researchers questions as they work."* For the next hour, science class middle-schoolers see, converse, and interact with Raptor Center staff scientists in real time as they go about their jobs at the facility.

In 1994, the Bell Museum launched the first episode of *Bell LIVE!*, a series of annual "electronic field trips" that were telecast live via satellite into classrooms across Minnesota and the nation. *Bell LIVE!* was modeled after the popular JASON Project hosted at the Bell Museum from 1990 to 2003. The JASON Project, the first of its kind, connected classrooms electronically and interactively to scientists and their research throughout the Americas. *Bell LIVE!* used a similar format but featured active research of University of Minnesota scientists working in Minnesota's ecosystems.

Raptors LIVE! was followed by *World of Wolves, Fire and the Forest, On the Prairie,* and *Great Lakes: A Superior Adventure.* Each fall during the project's seven-year run, more than eight thousand middle-school students connected and interacted remotely—live—with real scientists as they conducted their research.

But that 1994 launch represented something else. After a twenty-year decline as a venue for cutting-edge delivery of entertaining, science-based, natural history education, the Bell Museum was back. In fact, the revival of the museum's in-house film and media function was part of the museum's long tradition of innovation and leadership in the use of film to further the understanding of wildlife, natural history and environmental topics.

Film production at the Bell Museum began in the early 1900s with the Bell Museum's director Thomas Sadler Roberts, who was among the first to use film to capture wildlife behavior. Using a boxy, hand-cranked 35 mm movie camera, Roberts recognized that the new technology was an engaging way to reach a bigger audience, and a powerful way to build an awareness of the natural world. Beginning in the late 1920s, Walter Breckenridge expanded on Roberts's earlier work, turning out feature-length 16 mm movies on a range of natural history topics from Minnesota to the far-off Arctic.

Students studied stream life in the Bell LIVE! *Aquatic Adventure, 1998.*

Walter Breckenridge shoots footage for a movie about northern Minnesota wildlife, 1938.

In the 1940s, he hosted an immensely popular Sunday afternoon nature film series that regularly filled the four-hundred-seat Bell Museum auditorium to capacity. Breckenridge's fame as a film producer was recognized nationwide. In the early 1950s, he was contacted by Walt Disney Productions for use of his footage in Disney's *True-Life Adventures* series. The museum continued to capitalize on the public enthusiasm for such fare through the 1960s, before television nature shows and competition from other venues employing flashy, high-tech experiences began luring audiences away.

Following the success of *Bell LIVE!*, the museum formed Bell Museum Productions under the leadership of executive producer Barbara Coffin to carry this legacy forward into the twenty-first century. Coffin brought years of experience as an ecologist with the Minnesota Department of Natural Resources and The Nature Conservancy to her film productions. Her passion for Minnesota's natural environments made her well suited to lead the development of stories on the state's environmental history. Using a Ken Burns–style documentary format, the productions were intended to educate and inspire Minnesotans to connect with the natural world around them. On February 5, 2005, *Minnesota: A History of the Land* premiered on Twin Cities Public Television (TPT). The Emmy Award–winning four-part series (later expanded to five) on the history of the state's ecological landscapes showed the profound ways those landscapes shaped Minnesota's human history, even as those humans transformed that natural world.

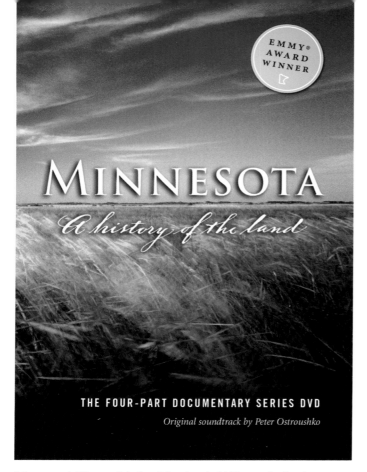

Minnesota: A History of the Land, *broadcast in 2005, was a landmark achievement.*

Barbara Coffin, executive producer, Scott Lanyon, museum director, and Robert Bruininks, University of Minnesota president, at the premiere of Minnesota: A History of the Land. *Photograph by Dan Engstrom.*

According to its Nielsen rating, more than five hundred thousand households across Minnesota tuned in to the first night's installment. The show that night made TPT the most viewed public station in the nation. Almost two decades later, the series is still broadcast on TPT two to three times each year and is used throughout Minnesota at the middle school, high school, and university levels. A companion curriculum and a teacher's guide have been made available free for download from the series' website, https://www.bellmuseum.umn.edu/history-of-the-land.

Bell Museum Productions went on to produce *Troubled Waters: A Mississippi River Story*. Broadcast in 2010 and based on solid science, the documentary explored the far-reaching and unintended consequences of agricultural runoff in the heartland and its impact downstream as far away as the Gulf of Mexico. The subject matter—a highly sensitive one throughout the nation's corn belt—created considerable controversy but also experienced a huge viewership. The show's content was discussed by audiences far larger and more diverse than the museum could ever have hoped for. Its use continues in college classrooms. At the University of Minnesota, it is a regular feature in a variety of curricula, including geology, agronomy, applied economics, and natural resource management.

Most recently, as the Bell Museum was about to move from the Minneapolis campus to its new location in St. Paul, the museum teamed up with TPT to produce *Windows to Nature: Minnesota's Dioramas*. This documentary tells the story of the Bell's renowned dioramas and their unprecedented move across town to a new museum building.

Although these long-format shows significantly raised the Bell's public profile, the museum was also using its growing production capacity to develop short-format videos to complement its museum-based exhibits and K-12 educational offerings. A series of ninety-second videos was developed and incorporated into museum-produced exhibitions showcasing the research of University of Minnesota scientists. The first of these, *Mysteries in the Mud: Climate Change in the Big Woods*, featured University geographer Bryan Shuman discussing his research and personal journey as a scientist. Other video-enhanced exhibits featured University researchers on topics as

Evaluation of the exhibit Mysteries in the Mud showed that a simple change in how a video program was presented made a big difference in helping visitors understand the process of science.

varied as microbiology, sustainable housing, honeybees, and astrophysics. The success of this dynamic storytelling format was recognized, and these video exhibits became prototypes for the Discovery Stations featuring University scientists in the museum's new building.

Minnesota Moose, a video-exhibit module produced by Bell Museum Productions in 2014, featured a touch-screen interactive next to the museum's beloved moose diorama. Three short videos highlighted University wildlife biologist James Forester's research into the decline of moose populations in Minnesota due to changing climate, disease, and moose–timber wolf interactions. Not only did it enliven the existing diorama experience, the video interactive served as a design prototype for the touch-screen interactives installed at each large habitat diorama in the new museum—a feature, it turns out, that is a favorite of all, from six- to sixty-six-year olds.

In 2018, digital production at the museum took another technological leap forward with the opening of the facility's state-of-the-art, 120-seat planetarium. Planetarium shows probing the mysteries of the cosmos through storytelling are enhanced by projected images on the overhead sky dome and include a live program in which a museum presenter responds to questions from the audience. Linked by powerful new communication systems to NASA and other data sources, presenters are able to deepen their responses to audience questions in real time.

The production of motion pictures at the Bell Museum continues. While the format may have changed with the development of new technologies, the long-standing tradition of movie making as an active program at the Bell Museum goes on. What has not changed is the museum's objective: to encourage the public to see science and nature in new ways and to explore human connections to the natural world through the roving lens of a camera.

From T. S. Roberts's hand-cranked films of the 1920s, to W. J. Breckenridge's Sunday afternoon nature movies of the 1940s through the 1960s, to the storytelling of research scientists of *Bell LIVE!*, to the exploration of Minnesota's place in the universe, the goal has been the same. As former Bell Museum director Scott Lanyon says in the closing narration of the fourth episode of *Minnesota: A History of the Land*: "We have an amazing planet that we live on, and we are, whether we like it or not, the stewards for that planet."

Minnesota in the Cosmos was the first original planetarium program produced by the museum for the new building.

Honeybees on the Roof

Sweetening Science Education

Honeybees are valuable educational partners. They are fascinating animals—much more complex than most people realize—and their critical role in human food production is too often invisible. —Kevin Williams, Bell Museum curator of public education

When Kevin Williams set up a collection of beehives on the roof of the Bell Museum, he never dreamed that his idea would become a model and catalyst for a wildly popular kind of science program embraced by the Minneapolis and St. Paul school districts. Williams was a curator of education at the Bell Museum in the 1990s and was an ardent hobby beekeeper. It occurred to him that these charismatic and intriguing insects might have great potential for educational programming. The architectural design of the museum's Church Street building included a twelve-foot-wide roof ledge adjacent to the third-floor classrooms. It was a perfect perch for beehives, and a great vantage point for viewing hive activity.

His idea was to establish a honeybee science lab for elementary-age students—an on-site facility staffed by a museum educator. Suited up as a beekeeper, the educator would work with the hives located immediately outside the windows of the classroom. A hive frame busy with bees would be pulled from the hive box and shown to students, who were inches away but safe from potential bee stings, behind the glass of the classroom windows. The primary goals of the honeybee lab were to teach kids insect anatomy, to have them learn about honeybees as social insects, and to role-play how beekeepers extract honey.

The program proved prescient. In 2006, a Pennsylvania beekeeper who was overwintering his bees in Florida reported the mysterious disappearance of much of his stock. This

Kevin Williams, on the museum's roof, shows a frame of honeybees to students in the classroom inside.

In the honeybee science lab students study bee anatomy with microscopes.

report was soon followed by other similar reports by beekeepers throughout the mid-Atlantic states and the Pacific Northwest in 2007. This bee pandemic of sorts was quickly dubbed a honeybee colony collapse disorder. Beekeepers were experiencing colony losses of greater than 50 percent. As the crisis spread, Bell educators realized that an expansion of the honeybee science lab could illustrate for school groups not only an understanding of the biology of a social insect but also the critical importance of a pollinator such as the honeybee to human food production.

Plans to take the honeybee lab to another level were developed in partnership with the Minneapolis school district's science teachers responsible for advancing the national integrated STEM (science, technology, engineering, and math) curriculum at the local level. A half-day residency titled *Honeybees, Pollinators, and Food* was developed in 2011 for the fifth grade. The central feature of the lab was an immersive experience at the museum. In addition to a field trip to the museum, the program included a pre-visit classroom presentation by a Bell Museum educator, and the delivery of a classroom kit with pre-visit activities, all of which provided a foundation for the on-site museum experience.

This was classroom learning with real-life experiences, hands-on objects, and inquiry-based learning. Aimed at fifth graders, the residency gave kids the opportunity to engage with the groundbreaking research of the University's renowned honeybee expert and entomologist, Marla Spivak. They also gained an understanding of structure and function in living systems through the identification of flower parts and their functions by dissecting flowers, and an understanding of animal adaptations through study of bees as pollinators. Finally and importantly, the lab revealed a personal connection for each student—the link between loss of pollinators and its impact on human food production. The lab soon found itself in high demand, limited only by staffing and classroom space in the museum's longtime home on Church Street in Minneapolis.

For ten-to-twelve-year-olds, a science lab taught by a university student can be an unforgettable experience with potential for sparking an interest in science that could lead to postsecondary study and a career in science. A ten-minute video, *The Practice of Science*, was produced in-house. The video featured Spivak's personal story and introduced the kids to the methodical steps that constitute the scientific process. As a

University of Minnesota bee researcher Marla Spivak was featured in a video that was a key part of the honeybee science lab. Courtesy of Marla Spivak.

fifth-grade teacher from the Bryn Mawr Minneapolis School wrote in her evaluation, "Being on a college campus impressed the students. For some students it signals that they could eventually become a member of this college group. When you see someone like you, it makes a place more welcoming."

The kids ate it up. In fact, a formal evaluation of the program showed that all who participated, regardless of gender, ethnicity, or income group, achieved significant new comprehension—a startling and welcome outcome. When the museum's new facility opened in 2018, the popular honeybee lab continued. Funded by the 3M Foundation, the lab was modified and made available to all second graders in the St. Paul school district. The tradition of beehives located just outside the classroom windows has continued.

The program has lived up to its potential. As Joe Alfano, Minneapolis K-5 STEM curriculum coordinator and partner in the development of the honeybee science lab, said, "When students are encouraged to learn inside and outside of the classroom, the learning is richer and deeper. All students finished this project with an in-depth understanding and appreciation for not only honeybees but also the world around us—this is invaluable."

Honeybees are fascinating animals with complex social behaviors. Here the "queen" bee is marked with the number 15.

Widening the Inquiry

Bringing Together Ecology, Evolution, and Behavior

Those were some of the most intellectually exciting years of my life, and they prepared me well for teaching and research at a liberal arts college. —Robert Askins, PhD, 1982

In the decades after World War II, chemists and physicists, whose work was so important to the war effort, turned their new tools and techniques to solving some of the long-standing mysteries of biology. They discovered the double-helix structure of DNA, cracked the universal genetic code, and revealed the mechanisms of protein synthesis. These advances revolutionized our view of life, opening up possibilities for human manipulation, such as genetic engineering (along with a Pandora's box of ethical challenges). From a molecular point of view, all life followed the same basic biochemical processes. From this perspective, such traditional disciplines as botany and zoology seemed passé, and the collection-based studies of natural history museums were seen by some as no more than mere stamp collecting.

Harrison "Bud" Tordoff was director of the museum from 1970 to 1983. University of Minnesota Archives.

By the 1960s, such attitudes had consequences. Wonder chemicals, like DDT, PCBs, and various mercury compounds that were supposed to save the world from hunger, disease, and poverty were mysteriously being altered in the food chain and concentrated to toxic levels. Pollutants from industry and automobiles that engineers thought would be easily and safely dispersed in the atmosphere were instead being converted by sunlight into layers of smog that were choking American cities.

Meanwhile, radioactive elements produced in nuclear weapon tests on the other side of the world were showing up in dairy products and even in mothers' milk. These unintended and unpredicted consequences resulted from a failure to consider the ecological systems into which these new inventions were introduced—systems that could only be understood through the detailed study of organisms in their environment.

Rachel Carson's book *Silent Spring* brought a new awareness of the global impacts of modern society and stimulated a rebirth of interest in ecology, wildlife preservation, and nature appreciation. Some of the signature icons in this environmental era were the images brought back by NASA's Apollo moon missions. What captured the public's imagination were not the images of the moon but rather the views looking back at Earth. For the first time we saw the reality of our existence—that all humanity and all of life are contained on a tiny blue marble, floating alone in the infinite inky-black emptiness of space.

MAN ON A MISSION

In 1970, not long after the first Earth Day, and riding a wave of renewed interest in ecology, Harrison "Bud" Tordoff became the new director of the Bell Museum. A former World War II fighter pilot, ardent ornithologist, and accomplished wing shooter, Professor Tordoff had big plans for the Bell.

The biological sciences at the University of Minnesota were being reorganized, and a new department of ecology and behavioral biology had recently been formed. Tordoff was determined to upgrade the scientific research and graduate student teaching at the Bell Museum and make it competitive with the best university-based museums in the country. He was in the enviable position of being able to hire four new faculty that first year. His recruits would be both curators in the museum and professors in the ecology department. All were young, eager, and ready to add their distinct talents to the success of the museum. Kendall Corbin was an ornithologist who had studied at Cornell and Yale Universities. He was one of the first museum scientists to use the new tools of molecular biology to determine the evolutionary relationships among species. Elmer Birney, from the University of Kansas, was an expert in mammalogy and also worked with molecular biologists to study the evolution of hemoglobin and ascorbic acid (vitamin C) metabolism.

But Tordoff wanted the museum to also be a prominent center for field biology. To that end, he hired David Parmelee, a noted expert on Arctic birds, to direct both the Cedar Creek Natural History Area with its radio-tracking program and the Lake Itasca Forestry and Biological Field Station. Both stations would go on to play important roles in the education and research of the museum's graduate students. Arguably, Tordoff's most unusual selection was Phil Regal as curator of herpetology (the study of amphibians and reptiles). An expert in physiological ecology and evolution, Regal was the closest the museum would have to a nineteenth-century "natural philosopher." He would find a topic of interest (such as the evolution of feathers), study it in depth, then come out with a big paper that would often shake up long-held views on the subject.

These museum curators formed the nucleus for a new Evolution Study Group that brought together distinguished faculty from across the University. Frank Barnwell came from zoology, Bob Sloan from geology, Sam Kirkwood from biochemistry, and even Ulysses Seal, a nationally recognized researcher from the VA Hospital, joined the group. Malcolm Kottler, a history of science professor with a specialty in evolution, also moved his office into the museum. Ernst Mayr, one of the founders of modern evolutionary theory, came from Harvard University as a visiting professor. It was an exciting time when new discoveries from field biology, ecology, animal behavior, and molecular biology were coalescing into a new understanding of evolution.

Tordoff instructs a student in the art of making a bird "skin" for the museum collection.

Elmer Birney's mammalogy class included a field trip where students learned to survey small mammal populations and prepare museum specimens.

Phil Regal with reptile collections preserved in alcohol, 1974.

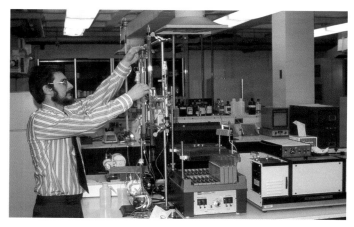
Kendall Corbin in his laboratory in the museum's basement, 1974.

But there was one more piece Tordoff needed to make the Bell Museum and ecology department nationally competitive: the program would have to attract top-tier graduate students. So Tordoff took over directorship of the graduate studies program. He required all applicants to pass muster by an admissions committee and then to pass an exam after their first year or two of classes. He wanted to weed out students who were unlikely to succeed. Doug Mock, who later became a distinguished professor at the University of Oklahoma, had arrived in 1969 to study animal behavior with Frank McKinney. Mock suddenly found himself facing the preliminary exam. "Of the six who took that battery of written exams, only two of us passed and the 'Reign of Terror' began," he said. "Suddenly, students began to get very serious about their graduate studies. Luckily, I managed to avoid the guillotine!"

CREATING A VIBRANT ENVIRONMENT

Besides setting high standards, Tordoff also sought to create a supportive environment for his students. Sometimes this was as simple as creating a tradition of timely breaks during the day when staff and students could get together. "Many of us well remember that Tordoff organized a coffee break in the morning and afternoon," recalled Francie Cuthbert, a graduate student in the late 1970s who later became a professor in the University's Department of Fisheries, Wildlife, and Conservation Biology. "It was a great place to get to know people and hear about what was going on." For David Blockstein, another graduate of the period, "Getting together every morning for coffee was a chance to listen to people and learn from them. Many of the students had international experience. Bob Askins and Daryl Karns had been in the Peace Corps. It opened my mind to think in a more global sense." Blockstein went on to help found the National Council for Science and the Environment.

In those days, the physical conditions in the museum were not great. Despite an addition added in the 1960s, graduate student offices were crammed into nooks between collection cases, exhibit workshops, and basement storage rooms. It was a cozy kind of chaos that, to hear those who lived through it, actually encouraged fellowship and cooperation.

"The most rewarding thing about the museum was the camaraderie among the grad students," remembers Bob Timm, now an emeritus professor of ecology and evolution at the University of Kansas. "Back in the mammalogy collection, Elmer [Birney] had us get some old furniture from the old Botany Building and create a 'bull pen' area. Each student had their own desk, but we had a group area for discussions."

Neil Bernstein doing field research in Antarctica, 1979–80. Courtesy of Neil Bernstein.

The students often learned as much from each other as they did from their professors or classes, the mark of a good graduate program. Many students wound up helping each other with their field research, naming other students as their best mentors. They would then go on to mentor new students. Joe Wunderle, who had a long career as a research scientist with U.S. Forest Service, remembers, "Joanna Burger was there and Doug Mock. Joanna, in particular, took me under her wing. From these students I learned how to be a good scientist. I had this role model and thought this is what you do—pass on the baton."

Students formed discussion groups that often met in the evening around a potluck dinner. They would discuss recent publications, share their research problems, and practice giving presentations. The best known of these groups was "The Fritterers," named because other graduate students in ecology thought they were frittering away their time doing intense research into the biology of a single species, when they should be looking at the larger ecological system.

Graduate students at the museum also had the unique opportunity to interact with the museum's public staff—educators, artists, and exhibit designers—something unusual

Graduate student David Bruggers studied raven behavior.

for a science-focused program. Graduate students were often hired to help develop exhibits, lead field trips, and prepare other public programs. The Bell even ran a series of popular informal evening classes at the museum. Teaching these classes helped to prime students for careers as professors. "Having all those things together, and the gallery with the art exhibitions, made it a very exciting environment," recalls Blockstein. "It gave us a much more complete experience as students to see the many dimensions of biology and the applications of biology. I learned how to communicate with an informal audience."

And just as Tordoff had hoped, the Bell Museum came to be seen as a center for developing expertise in field research. He and the curators encouraged their students to take intensive courses at the Lake Itasca Forestry and Biological Field Station and in Costa Rica at the Organization for Tropical Studies. Soon, these students were doing research in far-flung locales, such as Mexico, the Caribbean, and Antarctica. Neil Bernstein remembers arriving on campus as a new student in the fall of 1977, without a place to stay and without a job to help pay tuition. Within days, he was given a job and a place to live at Cedar Creek Natural History Area, and had signed up to leave in December on a research expedition to Antarctica.

Jane Packard, now an emeritus professor at Texas A&M University, came to the museum with a clear idea of what she wanted to do: study wolves. "I wanted to be the Jane Goodall of wolves," she said. "Of course this is not the way you are supposed to decide on your graduate studies. But at the museum, I felt part of a community. Nobody told me I couldn't do something." Packard went on to study with the nationally famous wolf researcher David Mech. She helped him set up his research station at Carlos Avery Wildlife Management Area. "The museum was merging the three major fields of inquiry—studying organismal biology and field biology by combining perspectives from ecology, evolution, and behavior. This was a most transformational experience," recalls Bob Askins, a retired biology professor at Connecticut College.

George Barrowclough, curator of birds at the American Museum of Natural History, came to the Bell in 1970 to study with Kendall Corbin. Barrowclough was one of the few graduate students at the time interested in studying bird evolution using molecular data. While the methods were crude compared to DNA sequencing, which would later transform this field, he could see even then that this kind of work would revolutionize the study of evolution. He considers his stint at the museum to have been a special time. "The five years I spent at the Bell were probably the happiest of my life," he said. "I was learning new things and discovering things on my own. I was introduced to the philosophical revolution in evolutionary and phylogenetic research taking place at that time."

Barrowclough could have spoken for many. It was the exciting churn of it all—the mix of stimulating faculty, top-flight graduate students, the rapidly breaking discoveries in ecology, animal behavior, and evolutionary biology, and the intensely collegial atmosphere that permeated the Bell—that created a time and space that many who experienced it would remember for the rest of their lives.

Graduate student Kevin Winker studies a radio-tagged wood thrush in the forests of the Tuxtla Mountains, Mexico. Courtesy of Kevin Winker.

Nature versus Nurture

Frank McKinney and the Evolution of Animal Behavior

In building an institution, there are sometimes those who wind up playing an outsized role. For the Bell Museum, one of those people was Frank McKinney.

From the outset, McKinney made a difference. When the British-born ornithologist and behavioral biologist joined the Bell Museum staff in 1964, his appointment brought with it a whole new dimension to the museum's natural history work. Frank's discipline was ethology—a fancy word for the study of the behavior of animals (including humans) in their natural habitats and how those behaviors likely evolved. Barely twenty years old, the science was one of many that arose after World War II.

McKinney learned his science from the men who invented it, the Austrian zoologist Konrad Lorenz and the Dutch biologist and ornithologist Nikolaas Tinbergen. Lorenz's groundbreaking work on the role of learning and instinct in the behavior of captive geese and other animals became the subject of his best-selling book *King Solomon's Ring*. From these two Nobel laureates, McKinney learned that the only way to understand what was happening in the nonhuman animal world was through long and patient observation to get to know a studied species intimately: what they did and when, and where they were and under what conditions. By

Frank McKinney photographs duck behavior at the Delta Waterfowl Research Station, circa 1960.

meticulously recording that information, it could be compared to the behavior of other closely related species. Common patterns of behavior could then be used not only to determine relatedness in species but also to point to a behavior's purpose and how it may have arisen and spread.

In 1959, McKinney immigrated to Canada to become assistant director of the Delta Waterfowl Research Station in Manitoba, a mecca for waterfowl biologists. It was there that he gained a reputation as an international authority on the social behavior of dabbling ducks. This is a large group of related species found throughout the world, with remarkably diverse behaviors.

At Delta he met John Tester, then a researcher at the Bell Museum, and the two became friends. When in 1963 McKinney decided to move on, Tester, who was about to launch a radio telemetry project at the University of Minnesota's Cedar Creek Natural History Area, invited him to join the project. McKinney jumped at the opportunity, joining the Bell Museum research staff that next year. Going on to become a member of the zoology teaching faculty, he subsequently became the Bell's curator of ethology.

McKinney constructed a set of specially designed flight pens at the Cedar Creek facility to house several species of captive ducks. He began studying their behavior and their responses to controlled experimental situations to test established theories. He augmented this work with field research, which took him to Alaska, South Africa, Australia, New Zealand, and the Bahamas, where he studied a range of duck species in the wild. These studies provided new insights into how ecological factors might have shaped such things as mating displays and territorial behavior.

The addition of McKinney was an important move for the Bell Museum in another respect: it put the museum and University among the vanguard of some of the most important research then being conducted in the field of behavior and evolution. In piecing together the selection pressures that had likely given rise to key behavior patterns, McKinney got to know the dabbling ducks like no one else.

In 1972, he started digging deeper and began looking at the possible conflicting objectives of males and females

McKinney weighs a duck during field research in Australia.

within the same species that would explain the differences he was finding in different species' territorial behavior, their mating strategies, and parental care systems. This turned out to be a hot topic, one that unleashed a flood of papers generated by a whole coterie of behaviorists who sought to show how evolution through natural selection could have given rise to a range of social behaviors, including altruism, infanticide, and differences in social systems.

Inevitably, these ideas were applied to human behavior. With the publication of biologist E. O. Wilson's book *Sociobiology: The New Synthesis* in 1975, a ferocious worldwide debate was unleashed over whether human behavior was largely a consequence of one's genes or one's environment—a nature versus nurture argument that for the next decade raged

Graduate student Gwen Brewer in the flight pens at Cedar Creek Natural History Area, 1980s. These pens were used to study the behavior of duck species from throughout the world under controlled conditions.

As a graduate student with McKinney, Doug Mock studied sibling competition among great egret chicks. Photograph by Doug Mock.

Frank McKinney was an internationally respected scientist and attracted highly talented graduate students to the Bell Museum.

not only in academia but in the popular press, in the halls of Congress, in faith communities, and in the streets.

McKinney was ideally positioned to pull these ideas together, putting them in proper perspective for his students. His classes were always packed. Indeed, he taught nine different formal courses and eight different graduate seminars, and supervised across his career some fifty undergraduate research projects on the behavior of a wide range of animals, from fish to reptiles to mammals.

But McKinney's greatest achievement was perhaps the influence he had on his many graduate students, a number of whom went on to have distinguished and groundbreaking careers themselves. He genuinely viewed and treated them as his peers, seeking to learn from them and their work, and was scrupulous about where and when credit should be given.

"Frank was unbelievably generous," recalled Sue Evarts, one of his students and research colleagues. "He was insistent about making sure we [students] be given credit for our own research." It was a sentiment echoed by many of his students.

"He was available to us for advice, but was hands-off," remembered Doug Mock, a biology professor at the University of Oklahoma. "Frank let me and most of his students design our own studies, make our own mistakes, and fix them as best we could. [And] he didn't slap his name onto papers we wrote. We were not livestock on his farm adding to his glory and to his professional record."

More than a mentor to his students, McKinney made them part of his family, frequently opening his home to them, where he and wife, Meryl, would host not only his students but their spouses and partners as well. Indeed, the close friendships that developed between his students produced more than simple camaraderie. Six of his students wound up marrying one another. The running joke among them was that McKinney seemed to be producing as many marriages as PhDs.

Mike Anderson, former chief scientist with the Institute for Wetland and Waterfowl Research at Ducks Unlimited remembered McKinney as "open, ever helpful, patient, a great listener, respectful, honest, but always encouraging and persistent about the things that mattered." One of those things that mattered was his students' work. Mock described McKinney as a "fierce" but fair editor. "One of the finest moments of my education," recalled Mock, "[was] when I realized he could remove 40 percent of the verbiage without loss of content!"

"Frank [would] return our manuscript drafts with red ink all over them, but encouraging side notes as well," recalled biologist Deborah Buitron. She and her husband, Gary Nuechterlein, are international authorities on grebe behavior and one of those married "McKinney" couples. "[But] when he said they were ready to submit to a journal, they were ready!" she said.

One of McKinney's trademark characteristics was the rigor he demanded from his students. He insisted that they continually question their own observations, conclusions, and what they read—including his own work. In return, he gave them his full support, cared about them personally, and became an advocate for their careers. So in demand was McKinney that by the time he retired, he had served on more than 150 PhD committees. Scientist, teacher, mentor, and human being, by so many measures, Frank McKinney was a most remarkable man.

As part of his doctoral research, Gary Nuechterlein built a floating blind in the shape of a muskrat mound. From within this blind, he could move slowly around the marsh, studying and filming the mating displays of western grebes. Photograph by Gary Nuechterlein.

Upland Sandpiper, *pen-and-ink illustration by Vera Ming Wong. Courtesy of Minnesota Department of Natural Resources.*

Minnesota's Rarest

Naming the State's Endangered and Threatened Flora and Fauna

Why concern ourselves with rare plants and animals? Aren't the common species of more importance to the long-term health and preservation of Minnesota's native habitats?

When Minnesota's first official list of rare plants and animals was published in January 1984, Minnesota author and naturalist Paul Gruchow reminded us of a list's elemental value: it puts us on notice. Such a list, Gruchow said, "has the eloquence of all lists and serves the same purposes. It names something and calls us to account for its consequences."

Common species, by their sheer numbers, might seem more important to the long-term health of an ecological landscape. But the rare ones are often the real indicators of conditions within that landscape. Recognition of the value of monitoring such species and the status of their populations has long been an important feature of the Bell Museum's work. In 1932, Thomas Sadler Roberts, looking back on the lost Minnesota landscape of his youth and contemplating the fate of a once common bird, said in his seminal book *Birds of Minnesota*: "To recite the history of the upland plover [upland sandpiper] is to tell a sad tale of the wanton destruction of a valuable and once-abundant bird that resulted in its almost complete extermination. . . . Now it is a question whether the remnant can be saved even with careful protection." Although the species did manage to survive the impact of market hunting in the late 1800s, then the loss of prairie habitat to large-scale agricultural conversion, its future still hangs in the balance due to changing rainfall patterns brought on by climate change.

When in 1981 the Minnesota Legislature updated the state's Endangered Species Act, the Bell Museum played a central role in the project. The act gave responsibility to the Minnesota Department of Natural Resources (DNR) to create a list of plants and animals classified as endangered, threatened, or special concern. To provide the scientific justification for what would be included on the list, the DNR was given the authority to appoint a thirty-member technical committee to advise it. As the DNR drew on experts from colleges and universities across Minnesota, it quickly became clear that the Bell Museum's curators and graduate students would play a critical role in the work of the committee.

Harrison "Bud" Tordoff, the director of the museum at that time, became the chair of the advisory committee and a member of the bird subcommittee. The museum's curator of mammals, Elmer Birney, and his graduate student Gerda Nordquist were made cochairs of the mammal subcommittee. Jim Underhill, the Bell's curator of fishes, was named chair of the fish committee, and Robert Bright, curator of mollusks, served on the invertebrate committee. Museum graduate student Daryl Karns, in the final years of his PhD research, provided expertise on amphibian and reptile subcommittees. Jeff Lang, who had completed his PhD at the Bell just five years earlier and was now professor of biology at University of North Dakota, was named chair of the amphibian and reptile subcommittee. The training and expertise of the museum scientists and their graduate students were a perfect fit for the task at hand.

The task they took on was monumental—evaluating the status of Minnesota's more than eighteen hundred vascular plants, six hundred vertebrate animals, and thousands of invertebrates and nonvascular plants. Knowledge of the distribution and abundance of Minnesota's native species was critical. This was particularly true for the vast majority of species that were not well known by the general public. For example, although

Minnesota anglers are very familiar with the state's game fish, most of the state's fish are small, nongame species. Underhill had just finished coauthoring the *Fishes of the Minnesota Region* in 1982. Earlier, in 1974, he had coauthored the third edition of *Northern Fishes with Special Reference to the Upper Mississippi River Valley*. Even earlier in the 1950s he had studied the distribution of small stream fishes in Minnesota for his PhD dissertation. There was no one better equipped to determine which species were in jeopardy.

Museum graduate student Jay Hatch seining in a stream in southwestern Minnesota as part of a survey of the Topeka shiner, a federally endangered minnow, 1980s.

Graduate student Sean Keogh with Higgins eye pearlymussel, listed as endangered on both federal and state official lists. It depends on deep free-flowing rivers with clean water, a habitat that is widely degraded by impoundments and diminished water quality. Like most other freshwater mussels, it is also threatened by the invasive zebra mussel. Photograph copyright Daguna Consulting.

Among the thousands of invertebrates that the committee considered for inclusion in the list were the state's freshwater mollusks. Mussels in particular were of special interest given their past economic importance to the pearl button industry and their renewed harvesting for the cultured pearl industry in China. The Bell's curator of mollusks, Robert Bright, was familiar with their historic significance and with the limited surveys that had been conducted since the initiation of Minnesota's Geological and Natural History Survey in 1872. Indeed, Bright was the most knowledgeable mollusk specialist in the state. The effort he and his team put in to identify all the state's rarest mussels convinced him that more needed to be done to monitor this group of organisms. Bright and his graduate students subsequently launched major surveys on the Pomme de Terre, Chippewa, Minnesota, and Zumbro Rivers. These became the foundation for a slew of additional survey and monitoring efforts across the state. Today, it is understood that mussels are the most endangered group of animals in Minnesota, the Midwest, and across the nation.

The work of the technical committee and DNR staff eventually led to the formal publication of the list in 1984. In 1988, the list was brought together in a book, *Minnesota's Endangered Flora and Fauna*, and made available to the public. The list was further updated and revised in 1996, and again in 2013. One illustrative example of the value in tracking rare species over time is Minnesota's native bats. In 1984, when Birney and graduate student Nordquist were working on the first list, they established a baseline of the statewide distribution and abundance of Minnesota's seven native bat species. Two of the seven species were placed on the 1984 list, but thirty years later in the revised Minnesota endangered species list of 2013, there were four on the list, principally victims of white-nose syndrome, caused by an invasive fungus.

As we advance further into the twenty-first century, the work continues, and undoubtedly the list will change yet again as climate change and human population continue to encroach on the natural places of the world. As ever, the scientists at land management agencies, universities, and the Bell Museum will be there to "call us to account for the consequences."

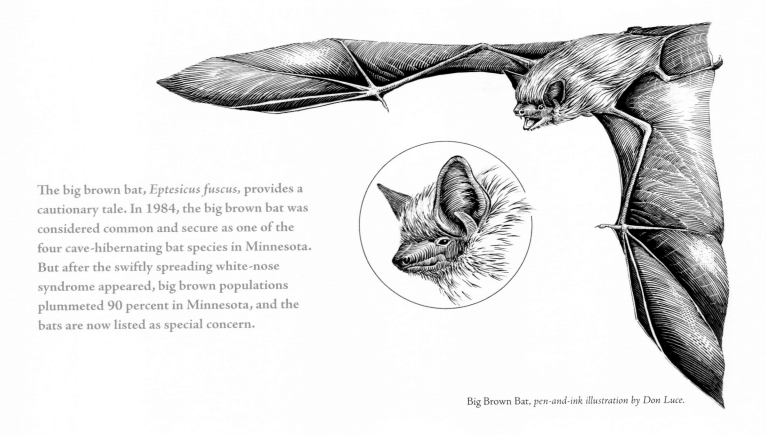

The big brown bat, *Eptesicus fuscus*, provides a cautionary tale. In 1984, the big brown bat was considered common and secure as one of the four cave-hibernating bat species in Minnesota. But after the swiftly spreading white-nose syndrome appeared, big brown populations plummeted 90 percent in Minnesota, and the bats are now listed as special concern.

Big Brown Bat, *pen-and-ink illustration by Don Luce.*

Graduate student Gerda Nordquist worked with Elmer Birney in the 1980s to survey bat populations in Minnesota. These data have proved particularly valuable now, when many bat species are threatened by white-nose syndrome, a fungal disease.

The word peregrine *means wanderer. Peregrine falcons range over every continent except Antarctica. Photograph by Greg Septon.*

Flight of the Peregrine

Bud Tordoff and the Return of an Endangered Species

Hopeless causes and the people who take them on have long been grist for epic tales and enduring legends. Take the story of the peregrine falcon, fierce predator of the skies, and the man who dedicated the last twenty-five years of his life to saving them from sure extinction.

The connection between Harrison "Bud" Tordoff and peregrines was something close to visceral. A lifelong hunter and former World War II fighter pilot, Bud once said that flying his P-51 Mustang was the closest he had ever come to experiencing what it must be like to be a peregrine. What moved him so was the bird's hunting prowess and spectacular aerial skills—its diving, twisting flights, its ability to turn a steep dive into a vertical ascent and, in the blink of an eye, to turn that ascent into a full loop the loop, flying upside down at the apex of its climb. And, of course, there is the bird's fabled attack dive, in which the bird often exceeds two hundred miles per hour as it closes in on its avian prey—the speed of its strike enough to kill its target in midair.

But for all its ferocity and skill, by 1960 the bird looked to be heading for the door. By then, peregrines were nearly gone from all of North America east of the Rockies and south of the Arctic. The victim of decades of habitat loss, shootings, egg collecting, and other human disturbances, the birds had already been in a steady decline. But after World War II and the widespread use of the pesticide DDT, the birds' numbers plummeted. The last confirmed nesting peregrine in Minnesota was recorded in 1962. By 1967, scientists had established that DDT was the principal cause of the birds' demise. The pesticide was officially banned in Canada in 1970, and in the United States in 1972. The following year the U.S. Congress passed the Endangered Species Act, and the peregrine was one of the first species listed.

While private falconers had been trying to breed the species for years, the first institutional effort to save the falcon began in 1970 at Cornell University with a captive breeding program designed by Tom Cade, a fellow falconer and professor at the university. With the nationwide ban imminent, the hope was that as residual levels of DDT declined, peregrine young raised in captivity could be reintroduced into the wild.

Cade's design, soon to be adapted nationwide, called for young peregrines around four weeks old to be placed in "hack" boxes—five-by-four-foot wooden containers about three feet high, equipped with a (removable) barred front to keep the young peregrines from falling out. The hack boxes would be placed on cliffs or atop towers and provisioned with food by staff. The bars would be removed after a couple of weeks, freeing the birds to roam their ledge or platform home and to test their wings. Within a couple more weeks, they would be able to fly, capturing butterflies or other small food items, and returning to their box at night. After about six weeks of flight time, the birds would be basically on their own.

Bud Tordoff with a peregrine ready to be released. Photograph by Steve Wilson.

DDT caused peregrines and other birds to lay thinner eggshells that crushed under the weight of incubating adults. Photograph by Derek Ratcliffe.

Patrick Redig and Bud Tordoff place captive-bred peregrine chicks into a hack box, 1980s.

In 1974, the first of Cade's birds were released in the East, and enough survived to encourage releases in the Midwest. In 1976 and 1977, eight birds were released from their historic nesting areas along the cliffs overlooking the Mississippi River valley near Maiden Rock and Nelson, Wisconsin. Two were promptly killed by great horned owls, one was found dead, and three simply disappeared. That was enough for the Cornell group, which decided against further Midwest releases.

Enter Bell Museum director Bud Tordoff. With no more peregrines likely from Cornell, he and Patrick Redig, director of the University of Minnesota's Raptor Center and a falconer, decided to begin their own peregrine reintroduction program. In 1982, they created the Midwest Peregrine Restoration Project to reestablish the birds across their former upper Midwest range. Their goal was forty territorial nesting pairs back on their historic sites along the Mississippi River and the north shore of Lake Superior. For Tordoff, a personal goal was to see a healthy population once again along the cliffs above the Mississippi in southern Minnesota, southwestern Wisconsin, and northern Iowa.

Using Redig's nationwide network of falconers and contracting with a peregrine breeding facility at the University of Saskatchewan, that first year the team released five falcons from a wooden tower on Weaver Dunes, a Nature Conservancy property near Kellogg, Minnesota, within sight of several historic cliff nest sites. Over the next couple of years, the team released thirty-three more birds. None proved successful at fledging young falcons. The naive birds were no match for marauding owls and raccoons. The Weaver Dunes site was abandoned after 1984 when it was discovered that surviving adult birds from the earlier releases were harassing the newly released falcons.

What to do? A chance suggestion provided an answer. As it happened, Tordoff sat on the board of The Nature Conservancy, and one of his fellow board members, a Minneapolis stockbroker aware of Tordoff's dilemma, suggested that they might have better luck releasing the birds from downtown buildings. His reasoning? Here was an abundant supply of food in the form of pigeons and other city birds, plenty of potential nest sites on building ledges and rooftops, and no predators. An additional benefit might be to excite public interest and involvement in the peregrine's recovery.

Tordoff liked the idea, and the team decided to give it a try. In 1985, six young peregrines were successfully released from a hack box atop the Multifoods Tower in downtown Minneapolis. Fifteen more were released the next year from the same site. In 1987, one of the released birds returned to nest. The hack box was replaced with a nest box, and the program was off and running.

Tordoff and Redig on the Multifoods Tower in downtown Minneapolis, 1985.

Young peregrines in a hack box.

Hack tower at Weaver Dunes near the Mississippi River.

MF-1, one of the first birds released in downtown Minneapolis, came back to nest on the building. Photograph by Patrick Redig.

The team subsequently began releasing the birds from other buildings in downtown Minneapolis and Rochester, and from cliffs along the north shore of Lake Superior. Word quickly spread, and in 1986 the recovery team began releasing birds in other states—Michigan first, then the next year Wisconsin, followed by the other midwestern states.

Tordoff, who by then had retired from the Bell Museum, became the public face of the falcon project. Redig managed the complex logistics that went with maintaining such an unusual supply chain and took care of the massive paperwork that accompanied working with an endangered species. Tordoff concentrated on raising money, enlisting the cooperation of corporate and institutional partners, banding and getting fledglings to the release sites, and raising public awareness about what they were trying to do. He was, as Redig put it, the "heart and soul" of the operation.

As demand rose from other states for falcon reintroductions, Tordoff and Redig decided to consolidate the operation at the University's Raptor Center, where the purchasing of birds could be centralized and where the fledglings could be vetted and banded, their health and pedigree recorded, and then resold. The project wound up supplying birds to thirteen midwestern states and the province of Ontario. A central database, established at the Raptor Center and maintained by Tordoff, enabled the team to track and keep tabs on the growing peregrine population.

The Midwest's peregrine population began rebounding rapidly. The success rate was so good that in 1991 Minnesota's urban release program was ended, and in 1996 it was stopped along Lake Superior. By 1999, the U.S. peregrine population had grown to a point that the bird was taken off the Endangered Species List. Birds successfully fledged from urban sites were now moving out and establishing nest sites beyond cities. In 2000, five nesting pairs were found on their historic nest sites on the cliffs along the Mississippi River. Now wise to the ways of owls and raccoons, all were successful, much to Tordoff's delight.

According to the Midwest Peregrine Society, which keeps track of the birds, as of 2018 there were 287 nesting pairs on permanent territories throughout the Midwest. These were

Fridge alights near her nest at Palisade Cliff on Lake Superior's north shore. In 1987, Fridge was one of the first city-bred birds to successfully return to natural cliff ledges. Photograph by Dudley Edmondson.

producing more than 520 young annually. Given the small but steady increase in peregrines over the past ten years, there are likely many more pairs today. Exact numbers are hard to come by, as tight budgets and state de-listings have caused significant reductions in monitoring. Minnesota is now home to around 70 nesting pairs, the most in the upper Midwest. In 1936 there were no more than 6 pairs nesting along the bluffs above Lake Superior. Today there are 26 pairs throughout the state's northeast, most on those same cliffs. Bud Tordoff, who died in 2008, lived to see his dream come true.

John James Audubon, Carolina Parakeets, *hand-colored engraving from* The Birds of America, *1826–38. Audubon strove to capture the motion and vitality of living birds in his art.*

Art and Natural History

The Evolution of a Legacy

In the early 1920s, Thomas Sadler Roberts got a long-distance phone call—an unusual event for the Bell Museum director, since long-distance calls in those days were rare and usually reserved for emergencies. The call was from his friend and fellow ornithologist Ruthven Deane in Chicago. Deane had a bone to pick with the good doctor.

"What business," Deane demanded, "does a William O. Winston of Minneapolis have buying a set of the original Audubon's *The Birds of America*?" Deane had a hobby of keeping track of original Audubon sets and would often stop by a certain rare book dealer in Chicago to peruse a set held by the shop's owner. But now that set was missing, sold to the unknown Winston. The *Audubon Folio*, as it was often called, was published between 1826 and 1838. It contains 435 life-sized images of birds engraved and colored by hand. It was bound in four volumes, each measuring almost thirty by forty inches, and each weighing about fifty pounds. It is now one of the world's largest and most expensive books.

As it turned out, Roberts was not only Winston's friend but also the family's physician. Still, the question puzzled him because he knew Winston was not a bird enthusiast. After the call, Roberts approached Winston and asked him why he had bought *The Audubon Folio*. Winston replied that Roberts was always talking about birds whenever he saw him. Since Winston had just built a summer home on Lake Minnetonka, he wanted a book to learn about the birds there. He had been told this was the best book there was on the subject, so he bought it. Each summer the four volumes were transported to Winston's lake home in a specially built cabinet. After Winston died in 1927, his family donated the *Folio* to the Bell Museum. Most people suspect that this was Winston's real intent all along. During their lifelong friendship, Winston knew of Roberts's love of birds and art, and there would have been no better way of recognizing it than with Audubon's magnificent *The Birds of America*.

In fact, art and science had always played a central role in Roberts's vision for his museum. Roberts had become the museum's director in 1919 during a national movement to transform museums from repositories of specimens to institutions where collections and research were combined with public education. He was particularly interested in the new science of ecology and in building public support for conservation. Dioramas combined art, science, and entertainment and were becoming powerful tools for popular education. The key to their success was their aesthetic appeal, which helped inspire viewers to care about the natural world.

Roberts knew his museum would not be able to compete in size with the large established institutions such as the American Museum of Natural History in New York. He was, however, determined to have his exhibits be their equal in quality. In building his new museum, he commissioned the best artists and taxidermists of the day to prepare the

Walter Weber, Sparrows, *watercolor from* The Birds of Minnesota, *1932. In the plates for* The Birds of Minnesota, *artists created accurate illustrations of the birds, with backgrounds that suggest their natural habitats.*

Francis Lee Jaques, Wings across the Sky, *oil on canvas, late 1930s. Jaques was a master at painting flocks of birds in flight.*

new dioramas. Four of the early diorama backgrounds were painted by Charles Corwin, one of the first artists to specialize in this artform. Bruce Horsfall, also a well-known wildlife artist of the period, painted two others. Edward Brewer, a Minnesota landscape painter and illustrator, painted the backgrounds of smaller dioramas. After construction of the new building in 1940, all the large and many of the smaller dioramas were designed and painted by famed diorama artist Francis Lee Jaques.

Dioramas were not the only place where art and science converged at the museum. One of Roberts's great dreams was to publish a comprehensive book on the bird life of Minnesota. He had been recording bird observations since 1874, when he was sixteen. He had many correspondents throughout the state who sent him their records. The book was to be illustrated by his friend Louis Agassiz Fuertes. When Fuertes died in a tragic accident in 1927, Roberts turned to the Canadian artist Allan Brooks, who agreed to do about half of the needed plates. George Miksch Sutton, a student of Fuertes, did some, and the young illustrator and animal artist Walter Weber did many others. Walter Breckenridge, then the museum's preparator, did fourteen plates after Roberts sent him to study with Brooks in British Columbia. The artist to round out the team, ironically, was Francis Lee Jaques, the man whom Roberts had turned down in 1924. *The Birds of Minnesota*, completed in 1932, is still considered a landmark in ornithological publishing. The

Thomas Bewick, Raven, *wood engraving from* A History of British Birds, *1797–1804. One of the more unusual parts of the museum's art collection is a group of fourteen engraved blocks used to print this historic book.*

Bruno Liljefors, Foxes, *oil on canvas, 1886. In 1988, the Bell Museum was one of only two museums in the United States to host the Bruno Liljefors exhibition* In the Realm of the Wild. *The show included magnificent paintings, such as* Foxes *from the Gothenburg Museum of Art in Sweden. Courtesy of Göteborgs Konstmuseum.*

noted wildlife art scholar and critic David Lank states: "Perhaps the most fascinating book for students of bird art is *The Birds of Minnesota*. . . . It was this book which confirmed the superior work of Francis Lee Jaques."

The connection between the Bell Museum and art became more explicit in 1971, when the museum put together and exhibited one of the nation's first major contemporary wildlife art shows. The *American Natural History Art* show included more than 160 paintings, drawings, and sculptures from artists throughout North America. Following this exhibition, Florence Jaques (Francis Lee Jaques died in 1969) offered to donate all of the drawings and paintings she owned to the Bell along with some funding. These funds, together with donations from many friends and relatives, covered the costs of remodeling the museum's mezzanine area into a gallery to display Jaques art, as well as rotating exhibitions of other wildlife art. Sadly, Florence passed away before the gallery was finished. On August 28, 1972, the gallery opened with a major show of Florence's collection together with other Jaques works lent by many friends and museums. The gallery was dedicated to both Lee and Florence Jaques. The museum's collection of paintings has since been supplemented by many other donations. The Bell Museum now owns the definitive collection of Jaques art.

Since 1972, more than seventy exhibitions of natural history and wildlife art have been displayed at the Bell Museum. Thirty of these shows featured works by Jaques, five were traveling exhibitions, which have been displayed at thirty-five museums across the country, including the Smithsonian Institution, the American Museum of Natural History, the Denver Museum of Nature and Science, and the Royal Ontario Museum. Other Bell Museum original exhibitions have included *A Celebration of Bats: The Photography of Merlin Tuttle*, and displays of Minnesota Duck, Trout, and Pheasant Stamp designs. Exhibitions from other museums have also been presented, including *J. Fenwick Lansdowne Watercolors: Rails of the World*, *Birds of Prey: Louis Agassiz Fuertes*, and *Pressed on Paper: Fish Rubbings and Nature Prints*. During the 1970s and 1980s, interest in wildlife art burgeoned along with growing popular concern for the environment.

In 1988, the museum opened a new and much larger changing exhibition gallery with the breathtakingly powerful paintings of Bruno Liljefors of Sweden. Many scholars recognize Liljefors, who worked in the late nineteenth and early twentieth centuries, as the Western world's greatest painter of wildlife. In conjunction with the exhibition, the museum organized *Wildlife and Art*, an international conference that explored the role of wildlife in art from prehistory to the modern day. Following this effort, the major exhibition *Pioneers of Bird Illustration* was organized and featured selections from the museum's complete set of Audubon's original

Gary Moss, Family of Swans—Trumpeters, *oil on canvas, 1977.*

double-elephant folio, *The Birds of America*. The show also included prints by the early bird artists Mark Catesby, Thomas Bewick, and Alexander Wilson from the eighteenth and nineteenth centuries. In 1991, the museum hosted the spectacular exhibition *Photography of Jim Brandenburg*, and later his wildly popular *Chased by the Light* exhibition in 1999.

The American Museum of Wildlife Art was founded in 1987 by William and Byron Webster. These brothers had built the highly successful wildlife-art print business Wild Wings in Lake City, Minnesota. As the business grew, they had collected original works by noted wildlife artists, including Owen Gromme, J. N. "Ding" Darling, George Browne, and Ogden Pleissner. The small museum was first located in Frontenac, Minnesota, and then moved to Red Wing, but it had a hard time surviving on its own. In 1993, aware of the Bell Museum's long history of integrating science, nature, and art, the Websters approached the museum with an offer to merge their collection with that of the museum. For five years, Byron Webster joined the Bell Museum's staff to help with public programming. Among the first joint ventures was the creation of one of the most comprehensive wildlife art exhibitions, *Wildlife Art in America*, which combined the best images from the recently merged collections, along with outstanding works by both historical and contemporary artists borrowed from other museums and private collections throughout the country.

Since then, the museum has mounted several major exhibitions of Jaques art and many other shows (a select list of these exhibitions is included at the end of this book), such as *Beauty and Biology*, an exhibition of the art and science of butterflies and moths. Other exhibition highpoints include the exhibition of exquisite botanical art by Margaret Mee, from the Kew Gardens in England, and *Visions of Nature: The World of Walter Anderson*, works by a remarkable artist who spent years exploring the islands off the Gulf Coast of Mississippi.

In 2014, the Bell again turned its attention to John James Audubon. With a major grant from the Institute for Museum and Library Services, funds were available for the conservation of the Audubon collection. At the same time, a major exhibition, *Audubon and the Art of Birds*, was organized that surveyed the development of bird art and ornithology from the Renaissance to the present day. Audubon's remarkable life and his amazing life-size paintings of birds were the centerpiece of the exhibition. This became one of the Bell's most popular exhibitions, which then traveled to other museums across the nation.

The key to all these efforts has been the presentation of art that is inspired by a naturalist's inquiring perspective. The museum's goal has been to display art that documents the earth's threatened biological diversity, explores humanity's need to understand nature, and inspires in visitors a desire to protect our natural heritage.

Andrea Rich, Bittern, *multicolor wood block print, 2008. The museum's art collection includes artworks in a variety of mediums.*

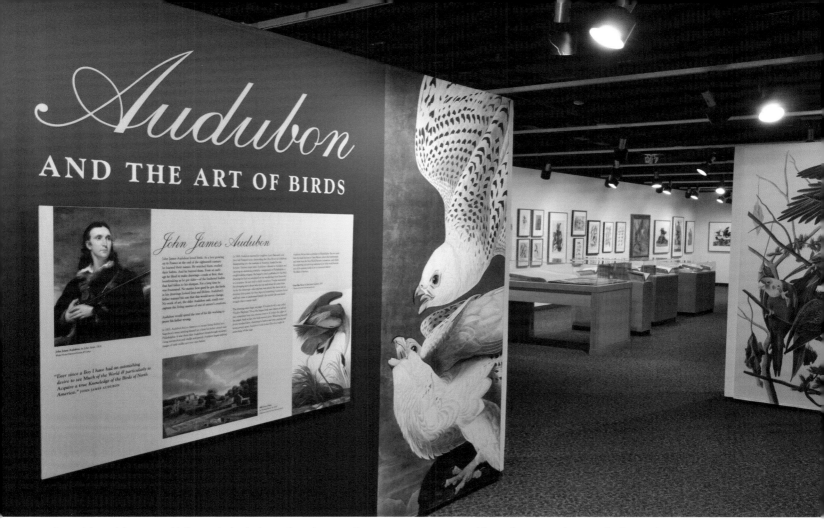

The Audubon exhibition assembled artworks that had never been shown together, giving visitors a once-in-a-lifetime chance to see the scope of bird-inspired art.

Encouraging the observation of nature is a key part of the museum's mission. In this activity visitors tried their hand at Audubon's drawing technique.

Susan Weller and Don Luce show members of the Board of Regents the Bien edition of the double-elephant folio displayed at the Audubon and the Art of Birds exhibit.

Science through the Lens of Art

Resident Artists at the Bell

Fans of natural history, popular culture, and the avant-garde can look forward to a unique combination of art installation and social happening at the Bell Museum Social. —University News Service, September 23, 2010

In the fall of 2010, the Bell took a chance and started a new evening program called the Bell Museum Social. It featured the artwork of a current museum resident artist, a presentation by a collaborating museum scientist, live music, food, and accompanying arts and crafts activities. Held periodically throughout the year, the Socials marked the beginning of the Resident Artist Research Project (RARP).

The Bell has had a long history of collaboration with artists, including artist-led workshops, gallery walkabouts, and group shows curated by artist-led organizations. But a formal, competitive artist resident program was something new for the museum. By the end of RARP's pilot years (2010–12), it was evident the Bell's gambit had paid off, as artists and collaborating scientific researchers using the museum's collections proved they could produce imaginative and stimulating work that engaged the public. Soon the McKnight Foundation became a long-term funding partner, enabling the Bell to offer resident artists generous stipends and to fund their art installations at the museum. Since then, the program has been open to artists from all disciplines, including writers, photographers, dancers, visual artists, puppeteers, musicians, and sculptors. The first public call for artists received an unexpectedly large response. Seventy Minnesota artists applied for four resident positions. This kind of robust response has continued to the present.

New works by RARP artists reflect the range of ideas and experiences possible in a Bell Museum residency—from Matthew Bakkom's interpretation of found photographs in the museum's archives depicting early scientific expeditions, to Olive Bieringa's experimental movement performance exploring mammal behavior, to interactive sound and light projections produced by Minneapolis Art on Wheels Collective that transformed the museum's iconic habitat dioramas.

During its first ten years, twenty-nine resident artists and eight showcase artists have participated in the program. Four artists are selected in each funding cycle for a twelve-to-fifteen-week residency. During the residency, the artists are given the opportunity to develop their artistic practice while exploring ways to interpret a particular scientific phenomenon. Two showcase artists are also selected as RARP alternates. Showcase artists present some of their existing work in a workshop or public exhibition. For example, in the show *Impact*, Miranda Brandon, a 2015–16 showcase artist, exhibited her powerful and evocative photographs of dead birds, casualties of collisions with glass-walled buildings. Michael Wilson (Makandwewininiwag Anishinaabeg), a 2019–20 showcase artist and electronic musician whose work is informed by his participation in the traditional Anishinaabe seasonal cycle, presented field recordings of animals and their environments in a hands-on sound-mixing activity to prompt exploration of climate change.

In 2011, Ali Momeni and Minneapolis Art on Wheels transformed the museum's dioramas with interactive projections.

Jeff Millikan, He Had Spent His Lifetime Gathering Eggs, *photograph, from his exhibition* Preserving/Memory, *2011. Courtesy of Jeff Millikan.*

Photograph by Miranda Brandon from her exhibition Impact: Birds in the Human-Built World, *2015. Photograph courtesy of Miranda Brandon.*

Resident artist Sonja Peterson discussed her cut-paper compositions at a Bell Social in 2013.

Donald Thomas, resident artist in 2019–20, works on his installation A Bird's Eye View of Climate Change *in the museum's basement workshop. Photograph by Joe Szurszewski.*

An important part of each residency is engagement with the public through interactive workshops, exhibitions, or other immersive participatory events. Resident artist Jeff Millikan created the exhibit *Collecting Memory: Photographs and Installations*, which was featured as an opening show in the new museum in July 2018. Puppeteer Christopher Lutter-Gardella, a 2017–18 resident artist, held workshops where together with the public he created *Star Man*, an articulated, eighteen-foot-tall marionette made of string lights woven through the sculpture's body, making it appear to be comprised of stars. Astronomer Carl Sagan's classic statement "We are made of star-stuff" inspired this work. Areca Roe, a 2011–12 resident artist, crowd-sourced images online from the public and placed them in "conversation" with scientific objects from the museum's collections for the exhibit *Freeze Frame: Capturing Nature in Winter*. As a finale to each residency, the resident artist's project becomes the focus of a night of art, science, fun, and entertainment at a Bell Museum Social.

The Resident Artist Research Project has added a new dimension to the museum's long history of integrating art expression with scientific discovery. The creative process at the heart of the artistic practice offers Bell Museum visitors the opportunity to "see" science through the lens of art. Artists unpackage the natural world in new and surprising ways with the potential to inspire curiosity and understanding in audiences of all ages and backgrounds.

A large, illuminated sculpture of the endangered rusty-patched bumblebee created by resident artist Anna Cerelia Battistini hangs in the museum's main gallery, 2020. Photograph by Joe Szurszewski.

Change Comes to the "Eternal" Museum

Temporary and Traveling Exhibits

Change is the only constant in life. —attributed to Heraclitus of Ephesus

Heraclitus to the contrary, museum galleries before the 1970s were pretty much static places. It would often take a decade or more to complete a new diorama hall or a dinosaur gallery. The Bell Museum's last large diorama, the Cascade River (Coniferous Forest), was completed in 1956, and the last medium-sized diorama, the Golden Eagle, was completed in 1968, twenty-eight years after the Church Street building was opened. When completed, these painstakingly developed exhibits were expected to last more or less unchanged forever.

For frequently returning visitors, this "eternal museum" provided the security that they could always return to savor their favorite painting, diorama, or artifact. But for most people, a visit to a museum was a rare thing, maybe three times in their lives: once as a child, once when they had children, and once when they had grandchildren. By the 1970s, natural history museums were searching for ways to increase attendance and encourage more frequent visitation. They were facing stiff competition from science centers and children's museums. For large museums, this was the beginning of the "blockbuster" age of special exhibitions that would attract long lines of visitors, such as *King Tut* and the *Chinese Warriors*.

The Bell Museum, like many smaller museums, also saw changing exhibitions as a way to encourage return visitation. But the 1940s-era building had only one small room available for changing displays, and this had been converted to a bookshop. In 1972, the mezzanine above the lobby was remodeled into the Jaques Gallery, a space devoted to showing changing exhibitions of nature-based art, often the work of local wildlife artists.

When Don Luce joined the exhibits staff in 1978, he was attracted by the museum's history of combining art and science. With a background in both art and science, and trained as a scientific illustrator, Luce saw that the museum's extensive collection of Jaques artworks was a largely untapped resource. He organized a Jaques exhibition that took a biographical approach. He used paintings, drawings, and field sketches to trace how Jaques, a backwoods farm boy, developed his skills and knowledge to become one of the world's most prominent nature artists. The show wove together information about environmental history, advances in science, and changing public perceptions about wildlife and the environment.

The exhibition *Francis Lee Jaques: Artist-Naturalist* opened in 1982 and then traveled to prominent museums across the country, including the Smithsonian and the American Museum of Natural History. The success of the Jaques exhibition demonstrated that the Bell Museum could create original changing exhibitions that were of interest to other museums. In the decades that followed, many more

Don Luce conducts a tour of the Jaques exhibition in 1982, the Bell Museum's first exhibit to travel nationwide.

exhibitions were created that explored the intersection of art, science, and nature. The Bell Museum developed a reputation for its exploration and display of natural history art.

Luce also saw the potential for changing exhibitions to highlight University scientific research and address current environmental issues. At the time, many of the museum's curators and graduate students were studying the peatlands of northern Minnesota in advance of a proposal to mine the peat and convert it to natural gas. These vast bog and fen communities were some of the state's most remote places. Researchers, funded by the Minnesota Department of Natural Resources, were urgently surveying the area's wildlife and their habitats to understand the ecology of the peatlands before a final decision was made on the mining proposal. Luce and the researchers traveled via helicopter to the center of one of the large bogs to extract a core through the deep layers of peat, for use in interpreting University research. The resulting exhibition, *For Peat's Sake: Researching Minnesota's Peatlands*, opened in 1984 and then traveled to venues around the state, including the State Capitol.

The peat exhibit was followed by many other exhibitions on topics related to University research, such as evolution, migratory bird conservation, aquatic invasive species, and endangered species. *Viewing Nature with Electrons* featured University research projects that used electron microscopes. To celebrate the twentieth anniversary of the College of Biological Sciences, the museum developed *Understanding Life's Connections*, which traced the study of life from the

The exhibition For Peat's Sake, *like many of the museum's traveling exhibits, was designed to be shown in non-museum settings, such as libraries, schools, and visitor centers. Here it is on display at the State Capitol in 1984.*

For Peat's Sake *featured Minnesota's extensive peatlands. Researchers had to use helicopters to access study sites.*

A visitor tries out the bait bucket interactive in the Exotic Aquatics exhibit.

The museum's traveling exhibits often presented current social issues, such as encouraging students of color to pursue careers in science.

Visitors at the exhibit Peregrine Falcon: The Return of an Endangered Species.

The exhibition Beauty and Biology: Butterflies and Moths in Art and Science.

DNA molecule to the global ecosystem. In 1992, the museum opened *Exotic Aquatics: The Biology of Alien Infestations* to help educate the public about the spread of invasive species. This potentially depressing topic was presented using cartoons and clever hands-on interactives. Three copies were made of this popular exhibit, and it traveled to more than twenty-seven venues in Minnesota and throughout the Great Lakes region. With funding from the U.S. Fish and Wildlife Service, the museum took a similar approach to presenting the topic of endangered species of the Midwest region.

When Kevin Williams joined the museum staff, he expanded the content of exhibitions to include controversial issues. *AIDS and Intimate Choices* opened in 1988 and was probably the first, and maybe the only, exhibition to address this difficult topic. *Many Faces in Science* opened the following year and featured scientists of color as role models for elementary-age students. One of the most popular exhibitions was *Tricking Fish: How and Why Lures Work*. The public's enthusiasm for fishing was an entry point to presenting fish biology. The Bell Museum's traveling exhibits program became known for its small- to medium-sized exhibits that were low cost and included simple interactive elements. Because of their size, many exhibits could be displayed in libraries, schools, visitor centers, and fairs as well as at other museums.

In 1983, when Don Gilbertson became museum director, he started a campaign to convert a long-vacant space on the museum's first floor into a modern exhibition gallery that could host temporary exhibitions. The West Gallery opened in 1988 with a major exhibition of wildlife art from Sweden, *In the Realm of the Wild: The Art of Bruno Liljefors of Sweden*. The show was followed by *The Language of Wood*, an exhibition of woodworking traditions from Finland. These exhibitions tapped into the large Scandinavian communities in Minnesota and opened the museum to new audiences. The new gallery also allowed the museum to develop larger original exhibitions. *Beauty and Biology: Butterflies and Moths in Art and Science* was developed in conjunction with University entomologists and drew upon the University's massive insect collection. The exhibition integrated art and science in new ways: it was organized around biological themes and included historic scientific illustrations as well as contemporary art, and even featured a live butterfly experience.

One of the museum's most enduring exhibitions was *Peregrine Falcon: The Return of an Endangered Species*. This exhibit traced efforts to restore these charismatic birds to the Midwest. The exhibition included many exquisite displays and small dioramas created by Bell Museum exhibit preparator Curt Hadland. The exhibition was so successful that the show was updated and redesigned in 2016, ten years after it was first created, and renamed *Peregrine Falcon: From Endangered Species to Urban Bird*.

From the 1970s to 2020, the Bell Museum presented more than two hundred temporary exhibitions. These included both rental shows as well as those developed in-house, and a select list of these exhibitions appears at the end of this book. At the program's peak, the museum offered sixteen different traveling exhibitions designed and produced by Bell staff. These exhibitions reached tens of thousands of visitors nationwide each year, greatly expanding the impact of the Bell Museum beyond its home base in the Twin Cities.

Visitors at the exhibition Sustainable Shelter *learn how the size and energy consumption of American homes have increased over time.*

The exhibit Hungry Planet *explored the diversity of foods available to cultures around the world and explained how diet choices impact health and the environment.*

Rediscovering the Collections

1980s–2022

Collections Offer Clues to Environmental Challenges

Off a dimly lit corridor deep below the glass-and-brick building that houses the University of Minnesota's ecology department, there is a locked, heavy metal door. Behind it lies a treasure of irreplaceable worth: a trove of skins, beaks, bones, eggs, and other specimens that are part of the natural history collections of the Bell Museum. Housed in several of the University's science buildings, the collections, along with those at universities and museums around the world, constitute our sole permanent record of Earth's known biological diversity. These collections are critically important to biological systematists, who curate, maintain, and use them to make order of the biodiversity around us. Together, the collections are the baseline material upon which the work of researchers throughout the biological sciences relies.

Curator of genetic resources Keith Barker shows a visitor the museum's bird collections from Central America. Photograph by Joe Szurszewski.

The Bell's collections are substantial: over forty-five thousand bird specimens, approximately forty-one thousand lots of fish specimens, twenty thousand reptile and amphibian specimens, some nineteen thousand mammal specimens, more than eighteen thousand lots of mollusks and crustaceans, and around nine hundred thousand plant, fungi and algae specimens—over one million specimens in all, and that is not counting the University's insect collection of over three million specimens.

Pass through that metal door under the Ecology Building, and you find yourself in the bird and mammal collection, a large brightly lit room filled with banks of high, gray-and-white metal cabinets. A fabulous menagerie of creatures that could never have existed together in the real world appears to cover every available surface. On a nearby table, a stuffed emperor penguin stands at attention in a glass box. Overhead a contemplative stuffed crow surveys an unlikely crowd of feathered colleagues, among them a pair of golden eagles, a cormorant, a pheasant, a toucan, and an albino ruffed grouse. Atop one of the far cabinets pokes the rear of a lion. Along the top of another bank of cabinets marches a row of small glass-fronted dioramas, while off along a far wall stands a small forest of antlers and horns hung from a long fence-like rack. Ringing the room and gazing down on it all are the heads of moose, cape buffalo, bighorn sheep, white-tailed deer, and a variety of other horned ungulates from around the world.

But the real treasure is in the cabinets. Behind their metal doors on the avian side of the room are stacked drawers of cotton-stuffed bird skins: meadowlarks, orioles, wrens, hawks, and owls, among others. Organized by species, a drawer will contain row upon row of individuals of that species. In the cabinets on the mammal side of the room, it is the same treatment for mice, voles, shrews, squirrels, and other small mammals. Elsewhere, other cabinets contain skeletons, eggs and nests, the skulls and bones of both birds and larger mammals, the lower jaw of the extinct great auk, and the leg bone of the extinct New Zealand moa. One floor down are the fish, reptile, and amphibian collections. Gray multitiered metal shelves, running the length of whitewashed rooms the size of single-car garages, are crammed with labeled jars of varying

Graduate student Kory Evans prepares specimens in the fish collection lab. Photograph by Joe Szurszewski.

sizes. In them float salamanders, frogs, snakes, and a seemingly endless variety of fish, some containing whole schools. There is enough combustible ethyl alcohol in these containers to warrant an escape hatch built into each room.

Spread across several rooms on the top floor of the adjacent Biological Sciences Building is the museum's herbarium. Here can be found more metal cabinets containing shelves of pressed plants laid out neatly on stacks of heavy paper, one specimen to a sheet. In another part of the room are rows of cardboard boxes containing lichen-bearing rocks. Elsewhere, a set of wooden filing drawers, not unlike an old library card catalog, contains blocks of wood from a range of tree species. Another cabinet contains fungi with their plant hosts on sheets, and dried mushrooms in boxes. And yet another cabinet contains dried mosses, and in a broom closet–sized room are metal cabinets containing dry algae.

This is the world of the curators.

When a researcher adds a new specimen to a collection, the date and location from which it was collected are recorded. These are the essential data, but other information, such as weight, sex, and reproductive status, is recorded whenever possible. This information is carefully written on the specimen label as well as entered into an electronic database. Records go back as far as the year 1800. Once a specimen is in the collection, the real work begins as the new addition is compared to

Curator Andrew Simons with graduate students in the mussel collection. Freshwater mussels are one of the most endangered groups of organisms in North America. The museum's collections are an important record of their past abundance and distribution in Minnesota. Photograph by Joe Szurszewski.

A fish head that has been cleared and stained to show its skeleton for research on jaw structure and feeding adaptations. Photograph by Joe Szurszewski.

other specimens. This is where the variety of the living world gets documented.

"Biology is the study of variation," explains Sharon Jansa, curator of the Bell's mammal collection. "Life on Earth takes various forms, and we are trying to understand the nature of that variation. To do that, we study how organisms are similar and how they vary both above and below the species level." And that is where the collections come in.

Because every individual of a species varies, each specimen is a unique record of that variation. But which specimens are variants and which are new or separate species is often hard to tell. A central part of the curators' work, then, is looking for shared traits, physical and physiological characteristics, and even behavior. These can include the physical appearance of various parts of organisms, cell structure, mating behaviors, habitat preferences, seed dispersal mechanisms, skeletal features, and, most recently, genetic makeup. It is important for the scientist to get this right. Biologists, whether they are studying the mating behavior of mallards or the neurology of newts, need to know the correct identity of the species they are studying. For them, a museum's collection is the ultimate reference.

These collections have value beyond the pursuit of science. In fact, they have proved immensely useful to us in more immediate, and sometimes unexpected, ways. For instance, in 1995, school kids on a field trip discovered hundreds of deformed frogs in a lake in the Minnesota River valley. The find led to an eruption of national headlines and launched an investigation into its cause. Deformities in frogs were hardly uncommon—they had been around as long as frogs had. What was unusual was the dramatic number of them. A major

Deformed frog from the museum collections. The Bell's collections indicate that these deformities were less common in the past. Photograph by Paul Nelson.

Photograph from the Bell Museum's website of the destroying angel, Amanita virosa, *that was used to correctly identify the mushroom that had poisoned a man in New York.*

suspect quickly became the pesticides that were applied to the surrounding farm fields. The Bell Museum happened to have a cache of more than three thousand leopard frogs collected during the late 1950s and early 1960s. An examination of them offered a tantalizing clue. Researchers went back to the same areas where the collections' specimens had been taken, and a new sampling of frogs was collected. Comparisons of the two groups showed a clear jump—a sixfold increase—in both deformities and their severity after just a few decades, a period during which there had been a significant increase in the use of farm chemicals. Further research has pointed the finger at something more complex than simple poisoning: it seems that the frogs' natural defenses against deformity-causing parasites that have always been in the environment are weakened by the new generation of farm chemicals, leading to more of the population being affected.

Another case occurred in upstate New York. In 2017, a man consumed a mushroom plucked from his lawn and became seriously ill. To correctly treat the patient, the hospital to which he was taken needed to know what kind of mushroom he had eaten. A slightly out-of-focus picture of the offending fungus was sent to an expert at the Poison Center in Syracuse for identification. It was initially determined to be a variety of common lawn mushroom that could sometimes wreak havoc with one's gastrointestinal system but only for one to three hours—and was hardly fatal. The patient's health continued to deteriorate, and the doctor at the center was increasingly unsure that the blurry image was indeed the mushroom it was thought to be. She turned to a professor of mycology at SUNY Cortland, who suspected the mushroom was something known as the destroying angel, a lethal fungus. To confirm his hypothesis, he logged into the University of Minnesota's online herbarium collection and matched the blurry image with the herbarium's image of the mushroom. Sure enough, it was the lethal species. Treatment was recommended on the spot, and the patient was saved, but only after spending several days in the hospital and after being transferred to another medical facility for further treatment.

Museum collections are indispensable to basic scientific research, but we should also remember the importance they can have in our everyday lives—and to future generations in ways that are impossible to predict today!

Swamp milkweed, Asclepias incarnata, *collected by Conway MacMillan in Crookston, Minnesota, August 1900.*

A Botanical Treasure

The University of Minnesota Herbarium

Not many people are aware that of the million-plus specimens in the Bell Museum's collections—birds, mammals, fish, amphibians, reptiles, mollusks, and crustaceans—the most extensive of them all are the nine hundred thousand, plants, algae, and fungi found in the museum's herbarium. The herbarium's collections, though not often seen by the public and rarely, if ever, mentioned in the press, have a history worth knowing.

In fact, the herbarium's beginnings in 1875 were as humble as one could get: a simple collection of plants in what was known as the General Museum, a room on the third floor of the University's Old Main building. This forerunner of the Bell Museum was the initial repository for all plants and animals collected by the Minnesota Geological and Natural History Survey.

But not until the University hired botanist Conway MacMillan in 1887 to instruct botany and to collect botanical specimens for the survey did the institution we know today, the University of Minnesota Herbarium, began to take shape. When a botany department was formalized in 1889, plant collections were then curated in the department's herbarium. Eventually, when the survey disbanded in 1916, its zoological collections went to the museum, and the plant collections that had remained with the survey were transferred to the botany department's herbarium. Over one hundred years would pass before the herbarium and museum were reunited in 1996.

Conway MacMillan came to the University of Minnesota in 1887 and was the first director of the University's herbarium. University of Minnesota Archives.

In those earliest days little was known about Minnesota's plant life. When MacMillan began surveying the state's flora, he did so by dividing the state into watersheds. He chose the Minnesota River valley watershed as his initial focus, finding it "peculiarly central in its location and remarkably interesting." By 1892, when MacMillan published his first botanical report, he had collected and prepared an astonishing twenty thousand specimens from that watershed.

The herbarium thrived and grew under MacMillan's leadership. In 1888, a year after his arrival, he oversaw the herbarium's first major accession: six thousand specimens from the Rocky Mountain states that had been collected by John Sandberg. A well-traveled Swedish immigrant, Sandberg had eventually settled in Minnesota. His collection, an impressive coup for the herbarium, cost the University $500. Although a significant investment at the time, it provided a solid foundation for further expansion of the herbarium.

MacMillan was relentless in his determination to build a first-class institution. Thousands more specimens followed the Sandberg acquisition, including those from MacMillan's Minnesota survey and collections from many noted botanists of the time.

According to his colleague Otto Rosendahl, MacMillan could be "rather temperamental." One of his students, Josephine Tilden, later wrote, "We all feared the man and cringed at being called on in class." But the professor also had a lighter side, which expressed itself during field exercises at the Minnesota Seaside Station on Vancouver Island, British Columbia. For example, his collecting expeditions included the comedic ritual worship of a mythological forest creature he dubbed the "Hodag."

By the time he left the University in 1906, MacMillan had recruited a cohort of botanists to Minnesota, who would expand the plant collections to include mosses, lichens, fungi, and specimens from far beyond Minnesota's borders. By 1914, the herbarium's collection was global, spanning North America, Europe, Australia, South America, and Africa. Despite early concerns over space, funding, and administrative indifference, it had acquired four hundred thousand specimens in less than twenty years. The herbarium as a quality scientific research collection had been launched.

AN INCUBATOR FOR SCIENTIFIC TALENT

Beyond its role as research institution and valuable repository for biological diversity, the herbarium was an incubator for scientific talent. This was particularly true for women seeking to break into science. Women could pursue botanical study much earlier than many other scientific fields. An article in the leading American journal *Science* in 1887 even argued that botany was not merely "suitable enough for young ladies" but also for "able-bodied and vigorous-brained young men." Early on, the herbarium helped launch the careers of four influential women: Eloise Butler, Margaret Oldenburg, Olga Lakela, and Josephine Tilden.

Olga Lakela works in the field. University of Minnesota Archives.

Margaret Oldenburg collected Arctic plants as an amateur botanist. University of Minnesota Archives.

Eloise Butler, an early "citizen scientist" and botany teacher, founded the first public wildflower garden in North America, located in Theodore Wirth Park in Minneapolis. After she moved to Minnesota from Indiana in 1874, Butler attended University botany classes and went on University expeditions. Her desire to have students experience plants in a natural setting led her to petition the Minneapolis Park Board, which set aside land for a botanical garden in 1907. She was the garden's first curator, and it was eventually named in her honor.

Like Butler, Margaret Oldenburg, a former University of Minnesota librarian, was a keen and accomplished amateur botanist. As an explorer, she spent time collecting in the Arctic during the 1940s and 1950s. The lichens, mosses, and plants she collected are among the herbarium's first from extreme latitudes.

Finnish-born Olga Lakela emigrated to northeastern Minnesota in 1906 and received her PhD from the botany department. In 1943, she was named the first head of the biology department at what was to become the Duluth campus of the University of Minnesota. Lakela was a prolific collector and became the foremost expert on the flora of northeastern Minnesota. The University's herbarium at Duluth is named in her honor.

Many other herbarium alumni have also gone on to have impressive careers. Generations of students trained by herbarium faculty have become leaders in the botanical fields of plant physiology, plant genetics, and plant ecology. University professor Otto Rosendahl, with student Frederic Butters, cataloged the Minnesota flora. Another of Rosendahl's PhD students, Arthur Cronquist, became one of the most influential botanists of the twentieth century and was director of the New York Botanical Garden.

MAPPING MINNESOTA FLORA

The herbarium has frequently partnered with botanists outside the University, in particular those with the Minnesota Department of Natural Resources (DNR). DNR botanist Welby Smith has been especially prodigious, contributing about 18 percent of the herbarium's nearly two hundred thousand Minnesota specimens, more than any other person. In 1979, Smith became the DNR's Natural Heritage Program botanist. At the time, he was the agency's only botanist and so earned the unofficial title of "state botanist." His first focus was documenting the state's rare and endangered plants. This included locating and describing rare plants to inform conservation and management decisions.

Smith's passion for collecting was inspired by an early mentor, Gerald Ownbey, then curator of the herbarium. Ownbey was concerned that Minnesota was falling behind its neighboring states, particularly Wisconsin, when it came to documenting plant diversity. He encouraged Smith to remedy the situation by collecting whenever possible. Ownbey, who died in 2010, likely would be proud of his protégé's achievement. During his forty-plus-year career, Smith has collected

Fred Butters at Minnesota Seaside Station, 1901. University of Minnesota Archives.

Professor Ernst Abbe, pictured with a herbarium specimen, did original work as a biochemical geneticist. University of Minnesota Archives.

specimens from all eighty-seven counties in the state, documenting new state records, gathering plants of all kinds, and filling gaps in the record of Minnesota flora. His work has become the foundation for mapping the state's plants, and his published guides to orchids, trees, shrubs, sedges, and rushes have enlarged the understanding of Minnesota's diverse plant life.

MERGING WITH THE MUSEUM

From the beginning, the herbarium has been an active research facility. By the middle of the twentieth century, it ranked twelfth (currently fifteenth) in the country in terms of specimen numbers. In those days, students and faculty collected and used plant specimens for plant classification, to document and describe new species, to verify species identification for studies of plant morphology and anatomy, and to document global plant diversity. Indeed, the growth of the herbarium and the botany department has mirrored the field's scientific advances, from simply describing flora and naming species, to parsing and dissecting the complex constituents of plant anatomy, to counting chromosomes.

But times and priorities change. By 1986, institutional support for natural history both at the University and throughout the nation had declined. In 1989, faculty and University administrators signaled a change in emphasis from specimen-based research to molecular approaches by changing the name of the Botany Department to the Department of Plant Biology. One hundred years after its founding, the herbarium was no longer at the center of the department.

To negotiate this difficult time, the department hired botanist Anita Cholewa as the herbarium's collections manager. Under her leadership, the organization of the collections dramatically improved. Specimens were rearranged to reflect the modern understanding of plant evolution, and taxonomic updates to individual specimens insured that the latest botanical names were available to the scientific community. However, even as the collection grew through ongoing contributions from the Minnesota Biological Survey, support for the herbarium continued to diminish.

Hard times often call for major rethinking, which sometimes leads to new clarity. From its beginning, the

Welby Smith, botanist for the Minnesota Department of Natural Resources, collected thousands of plants throughout Minnesota. Courtesy of Minnesota Department of Natural Resources.

Anita Cholewa, herbarium collection manager, oversaw major upgrades to the organization and taxonomy of the herbarium collections in the 1980s and 1990s.

University's natural history museum has been the repository for collected biological samples from the state and beyond. And that is what it continues to be today. Indeed, the Bell Museum has become an important University conduit to the public—a place where one can learn about the natural world. While the herbarium's position in the botany department may have made sense in 1889, that was not necessarily so in the late twentieth century.

In the end, the logic of a move to the Bell Museum was obvious. By 1996, the stage was set for transferring the management of the herbarium from the Department of Plant Biology to the Bell Museum. An agreement was reached for the museum to take on the herbarium's public and professional outreach and to seek external support for the management of the collection.

HERBARIUM FOR THE FUTURE

Today's herbarium curators exemplify what it means to serve the land grant mission—teaching, making new discoveries, preserving, maintaining, growing, and interpreting the biologically diverse collection. The value of their work, as researchers like Welby Smith can attest, continues to show how specimens and their attendant data, far from being outdated science, are fundamental to the science of tomorrow. As one herbarium curator puts it, "We keep collecting because nature keeps changing." Smith has likened herbaria to Arctic ice cores that preserve centuries of environmental change.

Herbarium specimens preserve patterns of diversity and distribution through time, helping us to understand past change and potential future scenarios. A specimen has permanence and can be studied even if its original location has changed beyond recognition. New species, new molecules, new genetic variation, and other kinds of information not yet imagined await discovery in the cabinets of the herbarium.

Tim Whitfeld joined the museum as herbarium collection manager in 2018. Photograph by Joe Szurszewski.

The DNA Revolution Comes to the Bell Museum

In 1992, the Bell Museum faced a major decision, one that would radically alter the nature of its research and the course of its own evolution as a modern natural history museum. By the 1980s, museum scientists were applying new tools from molecular biology to explore the age-old questions of natural history. What constitutes a species? How do new species form? How is one species related to another? What are the relationships among major groups of organisms? In essence, what is the pattern and pace of evolution? This field of study is called systematics.

Up until then, scientists would compare features of collected specimens that they could see and measure. But this line of inquiry was often subjective. Molecular tools offered a more objective way to compare species. With the development of DNA sequencing, scientists could now look at genetic variation in the very blueprint of life. DNA held the potential to revolutionize how we understand the evolutionary relationships of life.

As these developments unfolded, museum director Don Gilbertson decided to launch a campaign for an endowed chair to revitalize research at the Bell Museum. This was seen as a bold move for such a small department. However, the museum had many friends who loved birds, so it was decided that the endowment would support a new position for a prominent ornithologist. The chair was named in honor of former Bell Museum director Walter Breckenridge, who was well known throughout Minnesota as an advocate for both research and conservation.

The community response was immediate. Major gifts came from the James Ford Bell Foundation and noted Minnesota conservationist Wallace Dayton. But hundreds of donations also came in from individual supporters of the museum. The Minnesota Ornithologists' Union made contributions totaling over $25,000, the largest gift the organization had ever made. By spring 1990, sufficient funds had been raised to make the Breckenridge Chair of Ornithology a reality. The new chair would be the first new academic curator hired by the museum in more than twenty years.

A search committee was formed in 1991, chaired by museum curator Frank McKinney, who had an international reputation among ornithologists. After reviewing over forty applications and interviewing five finalists, the committee narrowed their decision to either a young, early-career researcher who was doing provocative work with DNA or a prominent ecologist who had written textbooks on ecology and was a member of the prestigious National Academy of Sciences.

Faculty in the Ecology, Evolution and Behavior Department (EEB) lobbied hard for the ecologist. But recognizing the need for the museum to change direction, the search committee unanimously selected Robert Zink, the young DNA researcher. Jim Curtsinger, a professor in EEB and one of the members of the search committee, remembers selecting Zink. As Curtsinger put it, "the museum, and the University, needed a researcher who knew how to use the new DNA sequencing technology. This was where the new big discoveries were going to be made."

In 1993, Bob Zink moved to the museum from Louisiana State University. He had received his PhD from the University of California, Berkeley, and then had been awarded a fellowship at the American Museum of Natural History. But he was really a hometown boy. He had grown up in south Minneapolis and had attended Roosevelt High School. As an undergraduate at the University of Minnesota, he had worked at the Bell Museum preparing scientific specimens and participating in research expeditions.

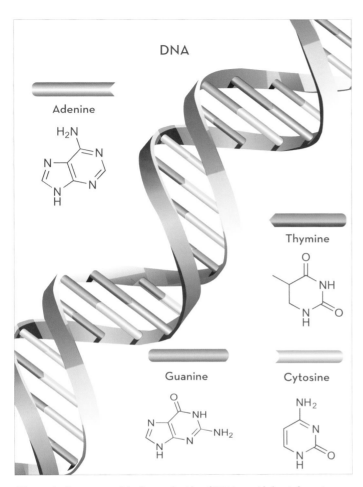

Changes in the sequence of the four nucleotides of DNA provide key information about the evolutionary relationships between living organisms. Copyright held by Shutterstock.

Robert Zink with some of his research subjects—fox sparrows. The plumage patterns of these sparrows vary greatly across their range.

Much of Zink's PhD research at Berkeley focused on geographic variation in fox sparrows. This bird ranges across the boreal forests of North America from Labrador to Alaska and south into the mountains of the West. Numerous subspecies of fox sparrows had been described over the years, but there was little consensus about whether this was one species or several. Zink's research combined fieldwork and classical morphological studies with cutting-edge analysis of the birds' molecular data. He concluded that fox sparrows should be divided into four species that are genetically distinct, despite some interbreeding between populations. He and other molecular systematists were proposing a whole new definition of what constitutes a species, challenging one of the foundational concepts of modern biology.

There were few things that Zink enjoyed more than ruffling feathers in the ornithological world. He and a Bell Museum graduate student, John Klicka, decided to challenge the long-held theory that Late Pleistocene glaciers had divided many bird species into eastern and western populations. The widely accepted theory was that Ice Age glaciers extending into the midcontinent had separated the populations of birds across the region, leading to the speciation of many east and west bird pairs. For example, there are the eastern and western bluebirds, the Baltimore and Bullock's orioles, the scarlet and

Baltimore and Bullock's orioles are known to hybridize in the Great Plains, where their ranges overlap. Photograph A by Carlos Roberto Chavarria; photograph B copyright held by Shutterstock.

western tanagers, and many others. Zink and Klicka compared the DNA from each pair and used it to estimate the date of speciation. The dates were much older than expected, predating the major glacial advances and thus refuting a theory presented in most textbooks.

In the 1990s, Zink got involved in a major controversy over the California gnatcatcher. The Southern California subspecies had lost 70 to 90 percent of its habitat to development. Its coastal range was in high demand for luxury home sites in the booming California real estate market. In 1993, the subspecies was listed as threatened under the Endangered Species Act. Conservationists looked to the gnatcatcher to bolster their efforts to preserve the few remaining tracts of open coastal sage scrub in Southern California. Into this controversy stepped Zink. He compared DNA samples from the California population to the far more numerous and non-threatened populations in Mexico's Baja California. He and his colleagues determined that the California population was *not* genetically distinct from those in Mexico, a finding that triggered a reaction that quickly became as much political as it was scientific. It is a controversy that remains contentious to this day.

But why stop at a tiny gnatcatcher when you might be able to gore the biggest sacred cow in all of biology—Darwin's finches? These rather drab birds of the Galapagos Islands had helped to inspire Darwin's theory of natural selection. Held up as a classic example of speciation, they are included in almost every biology textbook. Zink already knew that there was an ongoing dispute about how many species of Galapagos finches really existed, and even experts had a hard time telling one from another. Could an analysis of their DNA clarify the situation, or even better, maybe blow the whole textbook story out of the water?

Zink and graduate student Bailey McKay reviewed the morphology and genetics of the six generally accepted species of ground finches. They found insufficient divergence to support species-level ranks for these populations. Instead, they proposed that these populations had undergone many episodes of intense selection on bill and body size. But gene flow from birds flying from one island to another had prevented complete speciation. Zink and McKay then went on to give that process a grand new name, "Sisyphean evolution," after the Greek myth of Sisyphus, who was condemned to forever roll a rock to the top of a hill, only to have it tumble back down.

The California gnatcatcher is the focus of a conservation controversy. Photograph by Robert Hamilton.

This classic engraving of Darwin's finches illustrates the dramatic variation in bill size and shape among these birds of the Galapagos Islands. Illustration from Charles Darwin, The Voyage of the Beagle, *1845.*

Ruffled feathers or not, Zink has won some of ornithology's most prestigious awards. He was elected a fellow of the American Ornithologists' Union in 1988 and served as a councilor and vice president from 1992 to 1996. In 2005 he won the Brewster Award, given for the most meritorious body of work on birds of the Western Hemisphere published in the preceding ten years. In 2015, both McKay and Zink won the Katma Award from the Cooper Ornithological Society, given for work that could change the thinking about the biology of birds.

When asked about what theme has tied his work together, Zink said, "the study of evolution at and just below the species level." He was among the first to use DNA analysis for avian systematics and has helped make some astounding discoveries. Birds are one of the best-studied groups of organisms, and scientists have long thought there were about nine thousand species. DNA studies done by Zink and others have revealed a wealth of hidden diversity, suggesting there may be twice as many species as traditionally recognized.

Zink's imprint on the Bell Museum was major. When he came in 1993, he set up the first DNA sequencing lab for systematics at the University of Minnesota. His publishing record and prestige earned the museum support within the University and helped to attract outstanding new curators in mammalogy, ichthyology, herpetology, and botany.

After nearly twenty-five years with the Bell Museum, in 2015, Zink and his wife, Susan Weller, moved to the University of Nebraska, where he became professor, and Weller became director of the University of Nebraska State Museum.

In 2018, a new scientist was named to the Breckenridge Chair. Sushma Reddy is studying the diversity and molecular evolution of birds in India and Madagascar and their place on the tree of life. There are now seven curators at the Bell Museum actively using the research collections. All combine field and museum research with DNA analysis to make new discoveries about the amazing diversity of life.

Rethinking the Tree of Life

As buds give rise by growth to fresh buds, and these, if vigorous, branch out and overtop on all sides many a feebler branch, so by generation I believe it has been with the great Tree of Life, which fills with its dead and broken branches the crust of the earth, and covers the surface with its ever branching and beautiful ramifications. —Charles Darwin, The Origin of Species

For Charles Darwin, the Tree of Life was both a metaphor and a model for his great theory. Today it has become something else: a critically important research tool used to explore the evolution of life and to describe the relationships among all organisms, both living and extinct. It has become an enormous help to biologists who are seeking to better understand why modern organisms look and behave the way they do, and why they are found where they are.

Biological evolution, as we know it, is the process of inherited change in populations of organisms over many generations. As new species form from old ones, evolution can be pictured as a series of branching events. By selecting a group of species thought to be related by a common ancestor, evolutionary biologists can build what is known as a "phylogenetic tree." This is much like the familiar human family tree used by genealogists but based not on related people but on sets of shared traits. For generations of biologists, these traits were things that could be seen and measured, such as body form, skeletal structure, internal anatomy, and even behavior.

With the advent of many new tools from molecular biology, traits such as protein structures and DNA sequences can now provide new and powerful ways to explore the relationships of life. For evolutionary biologists like the Bell Museum's curators, DNA sequencing was "game changing." It gave unprecedented accuracy to the researcher's work. Trees became hypotheses that could be statistically tested, rather than unconfirmable conclusions based on comparisons and inference.

The development of rapid gene sequencing opened a floodgate of new information. In 2000, the National Science Foundation (NSF) launched the Tree of Life Project, an ambitious initiative to collect all known and current research

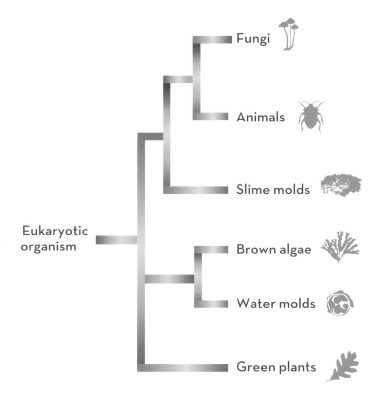

A phylogenetic tree illustrates how groups of organisms are related to each other based on shared characteristics. This tree shows how major groups of eukaryotes, organisms with a nucleus in their cells, are related. Based on both molecular and cellular data, fungi were found to be more closely related to animals than to plants. Illustration by Kari Finkler.

to build phylogenetic trees that accurately trace the path of evolution. Biologist Theodosius Dobzhansky famously said, "Nothing in biology makes sense except in the light of evolution." The trees resulting from the Tree of Life Project shine a bright light on all of biology, creating order out of the seeming chaos of life's diversity.

The Bell Museum's curators have been in the forefront of this research, and after the launch of the Tree of Life Project, the museum quickly became one of the top recipients of its support. Bell Museum curators have made many discoveries, some of which have refined our understanding of evolution, while other findings have upended long-held suppositions about the relatedness of species, and even of whole genera.

One of those scientists has been Bell Museum emeritus curator of fungi, David McLaughlin. Although now retired, he continues to partner with the Bell's curatorial staff in his research. McLaughlin has conducted groundbreaking research of this still little-understood group, which includes mushrooms, molds, yeasts, rusts and smuts, and aquatic basal fungi. This is a huge group of organisms, many of them of uncertain identity, and some of their lineages may go back as far as five hundred million years. There may be more than 1.5 million species in the group. No one knows, for only 150,000 have been described, yet they are critically important to the functioning of life on Earth. Among other things, fungi are principal actors in the decomposition of matter; they are essential mediators for proper functioning of plants and animals with which they are paired—including humans. As parasites, they are important regulators of plant and animal populations. Fungi make up about a quarter of the specimens in the Bell's herbarium.

Until fairly recently, the fungi were identified and classified according to their structures, most of them microscopic. In the past, their classification was based principally on sets of specific anatomical features, cell wall chemistry, and metabolic biochemistry. But now, as with all other lifeforms, gene sequencing has revolutionized what we thought we knew about these organisms. McLaughlin has been at the forefront of an international effort to understand and reassess the evolutionary history of fungi. As part of the NSF's Tree of Life Project, McLaughlin and four colleagues sequenced multiple genes from across all fungal groups to construct a phylogenetic tree robust enough to satisfy the leading authorities on all the groups of organisms that were sequenced. This new, updated classification has now been incorporated into major reference works and databases and was the subject of

David McLaughlin has been working to understand the amazing diversity of fungi since the 1970s.

a recent two-volume work on the earth's known fungi and fungi-like groups that also includes much new information on their evolution and ecology. "The exciting part of this collaboration is that we now have a fungal tree of life that for the first time shows reliably how all these organisms are related," said McLaughlin. "[That's] something scientists have sought since Darwin's *Origin of Species*!" One startling outcome of his and others' research has been the revelation that the fungi are closely related to animals and not to plants!

Another Bell scientist involved in working out the connectedness of life is Scott Lanyon. A former Bell Museum director, Lanyon and his postdoctoral student Keith Barker, now the Bell's curator of genetic resources, teamed up with other colleagues to review and bring up to date the evolutionary history and relationships of 832 species of New World

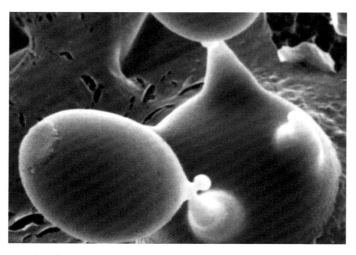
Identifying fungal species often depends on seeing their microscopic spore-bearing structures. Photograph by David McLaughlin.

Scott Lanyon is an expert on the evolution of blackbirds, a group that includes the large oropendolas of Central and South America. Photograph by Don Luce.

songbirds (warblers, sparrows, tanagers, cardinals, and blackbirds) by sequencing six commonly held genes. They used as their framework a renowned fifty-year-old foundational study conducted in 1946 by Ernst Mayr, one of the most renowned evolutionary biologists of his day. Mayr had taken on the monumental task of putting together a comprehensive analysis of the evolutionary origins of all New World bird species and the ecosystems in which they are found. Marshaling the best evidence available to him at that time and his own deep knowledge of evolution, Mayr had classified the majority of species by their continent of origin and reconstructed their worldwide dispersal.

Focusing on Mayr's subset of songbirds, Lanyon and Barker sequenced over 3.1 million base pairs of six commonly held genes—a task that kept their computers running for days. From these data, the scientists built a series of phylogenetic trees that enabled them to map the group's several lineages and secondary radiations with a precision that Mayr could have only dreamed of. In a testament to Mayr's skill, their tree analysis verified some of his major conclusions but also found other outcomes, which have opened up new lines of research.

Revealing the branchings of the tree of life is enormously important. The branchings help us better understand how today's organisms are related to one another. Knowing the structure of the tree of life helps us to understand key features of those species. In fact, efforts to fix a species' place in evolutionary time and space can often lead to unexpected discoveries. Take, for instance, the peculiarities of snake venom.

Snake venom is a notoriously complex and rapidly evolving trait—a big problem for medical researchers working on effective antivenom medicines. But what was driving this rapid evolution had long been a mystery. In the course of working out the evolutionary history of South American opossums, Sharon Jansa, the Bell's curator of mammals, stumbled onto one part of the answer. Jansa found that some tribes of opossums are predators of pit vipers and are immune to their venom. Coincidentally, one of the genes she was sequencing as part of her phylogenetic research was involved in the blood-clotting pathway. She noticed a curious thing. This gene showed unusually high rates of change over time in the venom-resistant opossums. Further research showed that this may be one contributor to the rapid evolution in the snake venom—under the selective pressure of their opossum

Curator of mammals Sharon Jansa with one of the opossums she studies.

predators, venom components could be evolving to counter the opossum's resistance. If true, this in turn would give rise to new, more effective versions of the resistant opossum gene in an endless biological "arms race" that may be contributing to the unusually rapid mutation of the snakes' venom.

The tree of life can also help us better understand events from ancient history. Andrew Simons is the Bell's curator of fishes. His investigation into the history and distributions of fish of the eastern United States' streams and rivers found evidence of the impact a long-vanished river had on the region's pre-Pleistocene fish populations. The Teays River was a major preglacial river that once flowed from Pennsylvania through Ohio and Illinois to the ancient Mississippi River.

Simons and his team set out to learn whether the distribution of fish in the central highlands of the United States could be explained by geological history. They did this by sampling fish species across their ranges and sequencing their mitochondrial DNA, constructing phylogenetic trees and mapping them to the terrain and known past geological events. In the course of this work, the researchers noticed that the populations of a particular species of fish—the northern hogsucker, a regionally widespread, bottom-feeding fish of fast streams—had an odd distribution pattern that could not be explained by the existing topography. The genetic makeup of hogsuckers sampled across the region showed a distinct similarity in the species' eastern and western populations—populations separated by hundreds of miles—but not with hogsucker populations sampled from streams in between. These hogsuckers had a distinctly different mitochondrial makeup.

"I thought, what is going on here?" said Simons. "The

pattern made no sense." But he did have a suspicion. "I told [one of my colleagues], 'You know, I think that's a signature of the Teays River.'" Geographers had long known the river was there, said Simons, but no one had ever found genetic evidence from the fish fauna to corroborate that.

They decided to find out. Correlating hogsucker DNA sampled across the old river's drainage system with the southern limits of the last glacial advance into the region, Simons and his team put together a picture that explained how the glacier's massive wall of ice could have effectively isolated the fish's eastern and western populations and how the subsequent final glacial retreat could explain the pattern they were seeing. They indeed were looking at the footprint of the old Teays River, and now they had the genetic evidence to prove it.

The tree of life work and the genomic research that underlies it continue at the Bell, enlarging what is known about the life around us. And where that work will take the museum's curators and what new and surprising things they will discover are sure to astonish.

Andrew Simons and graduate students Jacob Egge and Brett Nagle survey fish in Shoal Creek, Alabama.

Graduate student Hernan Vazquez loading samples into a thermal cycler to amplify DNA for sequencing.

Bell Museum Scientists on the Global Stage

Beginning in the mid-1970s, the biological sciences began undergoing profound change as field and lab workers studying biological diversity started to incorporate technologies developed by the biotech and medical industries. Researchers began using CT scans to make 3D measurements of variations in skeletons, gel electrophoresis to isolate and compare the proteins and DNA of different species, and modified histological processes used in human medicine to study the microscopic structure of the muscle and skeletal tissue of animals. Advances made in DNA and genomic research offered great promise but required large samples of purified DNA. When in 1985, a technique was developed for making millions of copies from only a small DNA sample, the floodgates were opened. And when in 2007, a new technology was developed that enabled rapid gene sequencing, the procedure went into overdrive. DNA sequencing output exploded.

The resulting flood of new information about the breadth and nature of Earth's biological diversity that these developments generated prompted the National Science Foundation in 2000 to commit major funding to further this research through the Tree of Life Project. The program's aim was to assemble a great "tree of life" by supporting multidisciplinary teams of researchers working out the evolutionary relationships and lines of descent for all life on Earth.

Because the Bell Museum had made the decision in 1993 to embrace just this kind of future, it was well-positioned to become a leader in this research—research that has taken Bell scientists to the far corners of the globe. The hiring of ornithologist Robert Zink that year was followed by the appointments of other scientists who were versed in these new technologies. Six new curators joined the Bell staff over the next thirteen years beginning in 1995 with the hiring of noted ornithologist Scott Lanyon as museum director. He was soon joined by ichthyologist Andrew Simons in 1998, botanist George Weiblen in 2001, mammalogist Sharon Jansa in 2002, herpetologist Ken Kozak in 2007, and Keith Barker, a genomic specialist, in 2008.

By the project's second decade, a new picture of life had emerged that, along with further advances in the technology, only served to raise new questions and complexities about the nature of nature. To meet the challenge, the Bell brought on a new cohort of researchers. Ya Yang, curator of plants, was hired in 2016. She represented a new generation of biological researcher eager to use the initial work of the Tree of Life Project to plunge deeper into the inner workings of evolution. As Yang explains, the revised picture generated by that early work involved using relatively few genes and short gene sequences. But it produced a genetic framework upon which a more detailed picture could be built. "Once you have the framework, then you can use the technology to study patterns

Botanist Ya Yang studies plant diversity with traditional and genomic techniques. Photograph by Joe Szurszewski.

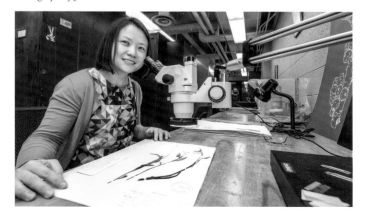

Breckenridge Chair of Ornithology and curator of birds Sushma Reddy in the bird collection. Photograph by Joe Szurszewski.

of complexities within genomes," she said. This can help researchers understand the processes that drive adaptations—to, in her words, "replay the evolutionary tape."

"We're now sequencing and reading ten to twenty thousand genes at a time depending on the plant species," she said. "We're discovering that things are a lot more complex than what a few genes can tell us. The tree of life turns out to be more a web of life!"

Yang is particularly interested in plants living in harsh environments, and her studies have taken her into the deserts of Mexico and the southwestern United States, to the mountains of Hawaii, and onto limestone outcrops in Florida. By studying adaptations that have arisen independently among different plant species in isolated, distinct, and geographically distant microhabitats, she hopes to get a better sense of what genes and gene sets are actively involved in particular adaptations.

Sushma Reddy is another scientist eager to build on the early tree of life work. Reddy was hired in 2018 to succeed Robert Zink as the Bell's Breckenridge Chair of Ornithology. She sees her research as global in scale, both geographically and phylogenetically. She is particularly drawn to working out the factors that have led to adaptive changes in groups of related bird species—a phenomenon called adaptive radiation. "A good portion of my work looks at the evolution of the entire radiation of modern birds," she said.

For Reddy, the research challenge now regarding bird species and their places on the tree of life is to resolve the myriad of complex and seemingly intractable relationship issues that have emerged from the early data. "These can't be solved by simply throwing *more* data at them," she says. "We have to address [these problems] through computational analysis, building different algorithms, thinking at the *genome* level . . . how genes evolve and how parts of genomes evolve."

Reddy has conducted extensive songbird surveys in such diverse places as southern India, Madagascar, Malawi, and Mozambique. By using data gleaned from combing through museum bird collections here and elsewhere, she has made some surprising discoveries in the evolution and biogeographic history of certain bird groups. One of the most remarkable examples of this is the story of the vangas, a group of medium-size songbirds endemic to Madagascar and wildly different in their appearance, particularly in the size and shapes of their bills. "Based on how they looked, they were thought to belong to many different families," said Reddy. "One was thought to be a babbler, another was thought to be a nuthatch, another a bulbul, another a batis [a shrike-like songbird]." Reddy's analysis showed that they all shared a single common ancestor that rapidly evolved into an array of morphologically diverse species—an adaptive radiation. Reddy wants to know why and how this happened. "[This knowledge] helps us to understand how much diversity there is in the world," she said.

Daniel Stanton, who curates the Bell Museum's collection of lichens and bryophytes, also sees the world as his research domain. Stanton was added to the museum's staff in 2019 for his expertise in mosses and lichens (the latter being made up of two different organisms—a filamentous fungi and its associated algae living in a symbiotic relationship). Stanton is especially interested in how these organisms interact with their environments and how they shape and even change those environments.

Like Ya Yang, he, too, sees the environmentally extreme areas of the world as excellent places to study how these organisms have evolved. Much of his research has been done at high elevations, on the side of volcanoes in Ecuador and on the coastal desert of Chile. "The Atacama Desert is the driest desert on earth," he said. "The organisms living there are at the limit of what multicellular life can tolerate. This is a place that can go fifteen to twenty years without rain. The few shrubs and cacti and a whole lot of lichen are sustained only by fog."

"These [life-forms] are tough," he said, and they are enormously important for the ecosystems they support. "They can grow on rock and they lay the groundwork for everything else that follows, not unlike what happened in [post-glacial] Minnesota ten thousand years ago." Particularly important are the mosses. "They profoundly affect carbon cycles, nutrient cycles, and water cycles." Ordinarily there is a certain predictability about the sequence of succession they trigger, he said. "But with climate change, the order in which [these successions] appear seems to be changing, [which] may lead to very unexpected outcomes. If you don't have the moss layer slowly accumulating nitrogen that takes years to develop, and instead you have weeds coming in, you may wind up with a completely different vegetative trajectory."

To prepare for the future, he said, "we need to make sure we understand the system"—an understanding that he is certain will be crucial in managing the coming changes to ecosystems around the world.

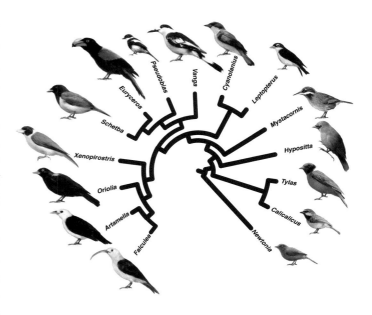

Phylogeny of the vangas of Madagascar showing dominant foraging strategy for each genus. From Reddy et al. 2012; illustrations of birds by Velizar Simeonovski.

Daniel Stanton sampling high Andean mosses on Cayambe volcano in Ecuador. Photograph by Evan Taylor Studios.

Biodiversity Research

Understanding Life's Threatened Diversity

The diversity of life on Earth and all its variations is mind-boggling. No one really knows how many species there are in the world, let alone what to call them. And it is possible that we will never know. Just as no one knows how fast those species are going extinct, or what the consequences of that loss will be. But there is a line of inquiry that aims to know just that. It is called "biodiversity science," and it makes up a large part of what the curators of the Bell Museum do. Besides eight to ten curators, usually a dozen or more graduate students and other researchers are using the collections at any one time in support of their biodiversity work.

George Weiblen is the museum's science director and curator of plants. He describes his work in broad terms. "All life shares common ancestry and a rich evolutionary history," he says. "We seek to uncover what is here now and how it came to be. We aim to interpret patterns of diversity and the processes that generated them."

There is a profound sense of urgency in this work, as the threats to biological diversity are increasing. Climate change, habitat loss, and a relentlessly expanding human population are pushing the planet's ecosystems to their limit—a trifecta that could drive species to extinction even before they become known to science. Documenting diversity and how it evolved has become one of the most essential roles of today's natural history museums—these "libraries of life."

Key to the researchers' work are identifying places on the planet that are unusually high in biological diversity—so-called biodiversity hot spots—and understanding how they got that way. The information gleaned can be used to develop management plans and policies to preserve areas that harbor a rich diversity of plants and animals—and maybe even to restore species in certain landscapes.

The questions biodiversity researchers ask tend to be big picture. Why and how do some parts of the planet harbor so many species while other regions do not? Why are there so many species in tropical forests? How are some kinds of organisms, such as invasive species, expanding their population sizes and geographic ranges in the face of dramatically changing environmental conditions, while others, such as amphibians, are rapidly declining and becoming threatened with extinction?

"Sure, it's interesting to count species and describe them, but these forms of life continue to evolve. They're dynamic entities," says Ken Kozak, the Bell's curator of amphibians and reptiles. "If we're going to have any hope of managing this legacy, we need to know something about the processes that generate and sustain diversity. We need to manage things in ways that enable species to resist or respond to change coming down the line."

The landscapes in which Bell curators and their graduate students work are scattered across the globe and are as varied as the questions these scientists ask and the work they do. But those scientists now have investigative tools to do that work that their predecessors of just a generation ago could only dream of—high-powered computers, techniques to examine parts of a species genome, advanced algorithms for inferring evolutionary history, and instant digital access to a wealth of past research.

The work of George Weiblen and Ken Kozak provides a case in point.

George Weiblen and rural subsistence farmers in Papua New Guinea established a community-based field station where citizen scientists participate in biodiversity research.

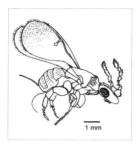

Fig trees are pollinated by microscopic wasps that depend on fig fruits for food and shelter. Illustration by George Weiblen.

Citizen scientist Brus Isua assisted Weiblen in biodiversity discovery. A new species, Brus's fig (Ficus brusii), was named in his honor.

George Weiblen with newly described fig species. Photograph by Joe Szurszewski.

GEORGE WEIBLEN, CURATOR OF PLANTS

George Weiblen's specialty is how plants interact with pollinators, parasites, and herbivores—especially if they have coevolved over time. He is interested in how interactions have influenced the formation of new species and their geographic distributions. His principal focus has been on wild fig trees in the mountainous rain forests of Papua New Guinea.

As a student, he was attracted to the complexity of fig pollination, requiring fertilization by certain kinds of wasps, and to how each of so many different fig species are pollinated by different wasps. But Weiblen's interest was also animated by tropical deforestation driven by industrial-scale logging and agriculture. As he put it, "I wanted to be close to it and do something about it." Papua New Guinea offered him all of that. It was the perfect place to study fig pollination. And it put him at ground zero in the struggle to preserve some of the world's most intact and least-known tropical forests.

His experience in New Guinea was life-changing and redirected his approach to research. Setting up a study area near the village of Ohu, about an hour from the nearest town, Madang, Weiblen quickly discovered that there were serious limitations to his Western education. "As a biologist my focus was on doing science. But I had to accept the fact that I was clueless and had no right to be there in the opinion of the Indigenous landowners," he said. Awed by their voluminous traditional knowledge, he partnered with local naturalists in conducting his research.

Since then, forest preservation in New Guinea has become a significant part of his work with Indigenous landowners. "There is cultural knowledge of these species that is at greater risk of extinction than the species themselves," said Weiblen. "Traditional ecological knowledge of biodiversity in places like New Guinea is disappearing at ten times the rate of species loss according to some estimates." And with that disappearance, he warns, goes an appreciation of nature that is every bit as valuable as the science.

He and colleagues founded a local nongovernmental organization in Madang, the New Guinea Binatang Research Center, to train people from local communities in biodiversity research and to provide critical technical support to professional scientists working in the region. Since then, Weiblen has accompanied other citizen scientists to spread the word to other communities in the region.

In 2001, Weiblen and colleagues were invited by leaders from the remote village of Wanang to study biodiversity on their land in exchange for education, employment, and access to health care. The scientists and their Indigenous collaborators attracted support from the U.S. National Science Foundation and John Swire & Sons, Ltd., a British multinational company doing business in Papua New Guinea, to establish an internationally recognized research station at Wanang.

Although balancing his community work and his scientific work is challenging, Weiblen still manages to do important biodiversity research. He was a principal scientist in an eight-year international study of rain forest plant–insect interactions that has begun to reveal the unbelievable complexity of a rain forest ecosystem. His research showed that a few hundred species of trees could support thousands of insect pests in a food web with tens of thousands of interconnections. Collecting, identifying, and preserving specimens are essential for making this type of discovery. In the course of his research, Weiblen has also named seven tree species and three insect species new to science.

KEN KOZAK, CURATOR OF AMPHIBIANS AND REPTILES

Ken Kozak's studies are concentrated in another global hot spot of diversity, the southern Appalachian Mountains of southeastern United States, an ancient landscape of never-glaciated wooded hills and valleys. "They've been eroding since the Paleozoic Era [some 500 million years ago]," said Kozak. "And the organisms that have come to live there have been through multiple climatic cycles [with] forested habitats expanding and contracting."

Kozak's specialty is herpetology—a group of animals that includes amphibians, like frogs and salamanders, reptiles, like snakes and lizards, turtles, and even crocodiles and alligators. But his main interest is salamanders, creatures that spend much of their time underground, under rocks, or in downed and rotting trees. These are all cool, moist environments that buffer salamanders against climate extremes. Although little noticed, salamanders can make up more biomass in the temperate forests of the Appalachian Mountains than all rodents and birds combined.

"They're particularly important in food webs," says Kozak. "So losing them is not only bad for diversity, but you're losing something that is really responsible for moving a lot of energy through the forest ecosystem." And that, he says, can have widespread effects on the healthy functioning of those systems.

And while salamanders' physiology and lifestyle make them particularly sensitive to environmental changes, they also, Kozak notes, have managed to persist, surviving the mass extinction of the Cretaceous-Tertiary. "That's not to say that that event didn't have massive impacts on diversity, but when the climatic conditions returned that were good, they were still there to fill and diversify in the resulting new niches that became available."

Many southern Appalachian species are ancient, something that is of particular interest to Kozak. "There are a lot of really old lineages there," he said. "Even though there have been a lot of climatic oscillations over time, the topographic complexity allows the species to not go extinct. They contract their range and persist in certain places. That results in more opportunities for species to accumulate in that region." Salamanders that require cooler temperatures to survive exemplify how the process works. During warming periods, they retreat to mountaintops and become isolated from other populations of their species. Over time, this isolation leads to speciation, the process by which new species arise, and an increase in salamander biodiversity.

In short, what Kozak and his colleagues have found is that climatic oscillations can for some species act as museums that buffer against extinction and also as engines of speciation. Landscapes such as the southern Appalachians and the mountain highlands of Central America—where he has also spent time—are ideal for preserving and producing new species. Kozak and his colleagues have found that a key element is time. The larger these ecosystems are and the longer they have persisted, the more species they will contain.

By gathering information about current climatic tolerances of present-day species and other traits key to their survival, and by mapping those traits on genealogical trees that extend back through time, Kozak and his colleagues have tracked changes in traits that have enabled scientists to reconstruct past conditions and past speciation events. "The earth has gone through a lot of warming cycles, yet biodiversity has persisted for really long periods of time," Kozak said, "so we can use that past as a window to the future, enabling us to hedge our bets as to where the best places are to put our [conservation] resources into managing such diversity." Kozak continues to be involved in international collaborations with scientists from several fields who are testing whether similar processes drive diversity patterns in other animal groups that inhabit mountainous regions.

Lungless salamanders, family Plethodontidae, can breathe through their moist skin. Photograph by Ken Kozak.

Typical salamander habitat in the southern Appalachian Mountains. Photograph by Ken Kozak.

Ken Kozak searches for arboreal salamanders that live in bromeliads in the cloud forests of Mexico.

One Hundred Years Later
Minnesota Updates Its Natural History Survey

What a difference a century makes. One hundred and fifteen years after the first Minnesota Geological and Natural History Survey, the legislature once again invested in a statewide survey but for vastly different ends. Rather than an inventory to expedite the extraction of the state's natural resources, the aim this time was for the purpose of conservation.

This second survey was begun as a collaboration between the Minnesota Department of Natural Resource's Natural Heritage Program and the Minnesota chapter of The Nature Conservancy, with support from the state's Environment and Natural Resources Trust Fund. Launched in 1987, the Minnesota Biological Survey (MBS) was the state's response to concerns that Minnesota's biodiversity was being threatened by real estate development and agricultural expansion. The idea was to build on prior work and to undertake a systematic survey of all eighty-seven counties in Minnesota to identify priority sites for conservation, and to generate data for meaningful statewide habitat comparisons. At the time, only Indiana and Michigan had similar survey programs, and no other state since then has undertaken a comparable effort.

The MBS began working in six prairie counties along Minnesota's western border and in Washington County, east of the Twin Cities. One of the first sites selected for survey was a forested area close to the St. Croix River in Scandia. The site looked promising on aerial photographs, but habitat details were unknown. So, with permission from the landowner, MBS ecologists visited and were impressed by the high biodiversity, intact forest, and dramatic landscape. Based on their survey, the site was recommended for protection and is now the Falls Creek Scientific and Natural Area (SNA).

Since 1987, MBS botanists and ecologists have continued to identify sites using aerial photography and ground surveys across the state. Thirty-three species previously not known to occur in Minnesota have been discovered. New populations of rare species have been documented, and thousands of acres of high-quality habitat have been located. In the process, a generation of field biologists has been trained. Precise information gained through the survey has allowed MBS staff to make recommendations for natural resource management to protect threatened and endangered species and to preserve habitat. The survey has compiled an unprecedented data set for informed decisions related to environmental review, watershed management planning, prairie restoration, technical guidance for landowners, and sustainable forestry.

As a result of the survey, over forty-five thousand specimens have been added to the Bell's scientific collections. These provide important documentation of the state's flora and fauna over the late twentieth and early twenty-first century, a period that is generally not well represented in natural history museum collections. The specimens are also an important component of the Minnesota Biodiversity Atlas, a web portal

Minnesota Biological Survey ecologists conduct a vegetation plot, known as a relevé. Courtesy of Minnesota Department of Natural Resources.

to records of Minnesota's plant and animal diversity. From diminutive and rare plants such as false mermaid and bog adder's mouth orchid, to favorite wildflowers like large-flowered trillium and showy lady's slipper, MBS specimens fill in gaps and allow researchers to track changes in the distribution of species through time. In this way, MBS operates in the same spirit as the original Minnesota Geological and Natural History Survey, which founded the scientific collections of the Bell Museum.

In addition to collections, many other tangible products have come from MBS work. One of the most significant is the mapping of native plant communities and rare species across the state. Thousands of vegetation plots (relevés) have been surveyed, continuing an effort started by University professor Edward Cushing some five decades ago. These plots are foundational to Minnesota's plant community classification system, which has guided natural resource management in the state since the 1990s. Survey data continue to support recommendations to government agencies and conservation organizations when parcels of land worthy of protection have been identified. Falls Creek SNA is just one example. All survey data are now available online in the Minnesota Biodiversity Atlas.

With the exception of a few areas in far northern Minnesota, the original county-by-county survey is mostly complete, but in many ways MBS is just getting started. What the survey has now documented constitutes the baseline data necessary to understand how life in Minnesota is adapting to changes resulting from human activities like habitat loss from development and other causes such as global warming and the presence of invasive species. Invertebrate surveys have been initiated, and surveys of aquatic plants, nongame fish, mosses, and lichens continue. Gaps are being filled with a focus on Minnesota's endangered prairie habitats and more intensive surveys of other special areas. Another effort focuses on long-term monitoring of vegetation to detect and measure change in the presence/absence and population levels of native plants. Collections will continue to flow into the Bell Museum, and more detail on Minnesota's plants, animals, prairies, forests, and wetlands will be added to our collective knowledge and available for future assessment and management of Minnesota's biodiversity.

Sneezeweed, Helenium autumnale, *collected 2004. Thousands of plant specimens collected by Minnesota Biological Survey botanists are stored in the University of Minnesota Herbarium at the Bell Museum.*

False mermaid, Floerkea proserpinacoides, *is one of many rare plant species documented by the survey. Photograph by Peter M. Dziuk.*

In 1994, the four-toed salamander was discovered for the first time in Minnesota by Minnesota Biological Survey scientists. Photograph by Andrew Herberg, Minnesota Department of Natural Resources.

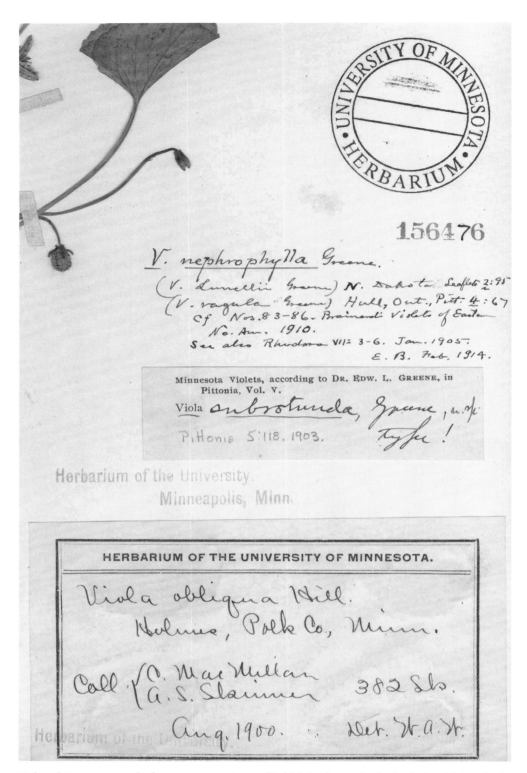

High-resolution scans are made of every museum specimen and its label. But data on the often handwritten labels need to be accurately entered into the digital database. This is an opportunity for volunteers to help in the Mapping Change project.

Collections Go Online

It might not be obvious, but a museum specimen is only as valuable as the quality of the information that goes with it. To wit: where it was collected and when, what it is, and who collected it and why.

The data recorded in museum catalogs and on labels are what give collections their scientific importance. Exquisite pen-and-ink handwriting on nineteenth-century specimens and the early adoption of typewritten or printed labels by twentieth-century collectors bear witness to this fact. A formal education in how to curate a natural history collection in the past necessarily required knowing how to negotiate one's way through card catalogs and other analog media. Should a specimen be misfiled or a taxonomic name misapplied, the error might go uncorrected for decades. The upshot? Curatorial practice tended to favor gatekeeping over accessibility.

When library card catalogs gave way to searchable databases in the 1970s, Bell Museum curators were not necessarily keen to relinquish control over their records to computers, then known as "business machines." Curator of lichens and professor Clifford Wetmore was a revolutionary in this regard. He was among the first in the United States to digitize an entire biological specimen collection, and he began doing so at a time when data were represented by tiny holes punched in card stock. The processing of records involved hand delivering a box of punch cards to the University of Minnesota's mainframe computer and crossing one's fingers in hope that a card would not jam the machine. Labels could be printed automatically with the touch of a button and without the inevitable errors introduced by a typist, typically some well-intentioned student with a stern curator lurking nearby.

Fast-forward to the 1980s and the advent of the desktop computer. Storage on punch cards and magnetic tape gave way to digital media and the power to manage an electronic catalog right in the heart of the collection. While a generation of curators grew accustomed to directly managing their data, an ever-changing variety of digital formats did not instill confidence that records would be safe if a curator ever lost his or her computer. Even worse, what if administrators concluded that the data alone would be a sufficient record and the University could save on costs by disposing of the specimens?

Historically, curators of the various Bell Museum collections tended to make their own choices and manage data in different formats. In the 1990s, director Scott Lanyon recognized the opportunity to unite curators around a common database platform. The Bell Museum became an early adopter of the natural history collections software called Specify at a time when desktop computers were being wired together into larger and larger networks. Pressure mounted to share data like never before on the "World Wide Web." The museum participated in federally funded efforts during the 2000s to make snapshots of collection records available to researchers through data portals operated by other museums.

A significant advance in the public accessibility of the Bell's collections occurred in 2016 when the museum formed a partnership with the University's Supercomputing Institute. Curator of genetic resources Keith Barker obtained support from the Legislative-Citizen Commission on Minnesota Resources of the state legislature to create the Minnesota Biodiversity Atlas, an online tool for searching, mapping, and visualizing the entire scientific record of the museum. This project made the museum among the first in the nation to operate a data portal offering simultaneous access to all kinds of biodiversity, from big bluestem grass to zebra fish.

A student makes a scan of a plant specimen in the herbarium. Photograph by Joe Szurszewski.

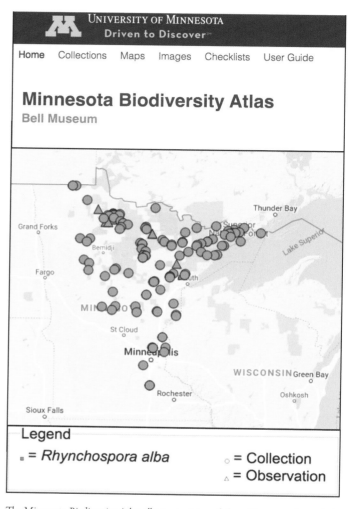

The Minnesota Biodiversity Atlas allows scientists and the public to see all records for a species and a map showing its distribution in Minnesota.

Thus began a new chapter for the museum in which virtual visitors could access the collections from anywhere at any time. Today, other state institutions, like the Minnesota Pollution Control Agency and the Minnesota Biological Survey, use the data found in the Minnesota Biodiversity Atlas, pushing it out to even larger biodiversity repositories around the world. "This is one of the greatest research and conservation resources at the University of Minnesota, and now it's the most accessible," said Barker.

Professionals are not the only ones who contribute to and use the Minnesota Biodiversity Atlas. Through the University's Mapping Change project, citizen scientists transcribe handwritten museum specimen labels, helping to build the digital contents of the atlas one specimen at a time. The project is all about making the scientific collections accessible for researchers and the public. All survey data are now available online in the Minnesota Biodiversity Atlas, both locally in Minnesota and worldwide to researchers wherever they might be located. As Barker puts it, now "with a few clicks, anyone from researchers, to nature enthusiasts, to educators can locate records of life in Minnesota and beyond."

Scientific collections, such as this tray of blue jay specimens, record the distribution of species as well as their natural variation. Photograph by Joe Szurszewski

A Museum for the Twenty-First Century

1990s–2022

Saving an Endangered Museum

Surviving and Thriving in a University Setting

After the vetoes, combined with the recession, it was a double whammy. The public had forgotten we existed, and staff morale was terrible. No one thought the project would ever come back. —Susan Weller, former Bell Museum director

Buffeted by conflicting institutional agendas and forces out of its control, the Bell Museum's rocky journey to its new home on the University of Minnesota's St. Paul campus reads like something out of Homer's *Odyssey*. Indeed, the Bell's struggles over the past forty years neatly illustrate the many difficulties faced by natural history museums within research-oriented universities across the country. The fortunes of universities and their museums began to decline in the 1980s when the national trend to reduce taxes led to severe reductions in state support. Repeated budget cuts created a cutthroat atmosphere on campuses, as academic departments within the institutions competed for dwindling dollars.

At the University of Minnesota, the Bell Museum was at particular risk. School groups made up much of the museum's audience, and many academics questioned whether the education of young children was central to the University's mission. On top of this, natural history was viewed as outdated, and many of the museum's curators were nearing retirement. And finally, the College of Biological Sciences, to whom the Bell Museum reported, did not view the museum's public programs as a priority.

When Don Gilbertson, an ecology professor, became director in 1983, succeeding Harrison Tordoff, he recognized that he had to improve the museum's situation. He formed an external advisory board, started a membership program, and instituted museum admission for the first time. He also started a quarterly newsletter, *IMPRINT*, which promoted the museum's exhibitions and education programs as well as highlighted its current natural science research and that of other University departments. But the erosion continued. By the time Gilbertson retired in 1990, museum faculty positions were not being replaced as curators retired.

When Elmer Birney, curator of mammalogy, took over from Gilbertson, he decided to set a new course for the museum. He helped to start a consortium of university-based natural history museums from around the country to discuss strategies to enhance their programs. Consortium members knew that most funding for scientific research came to the nation's universities. University museums, they reasoned, should be well positioned to interpret that research to the public. They also knew that by and large university museums were among the most outdated and underfunded museums in the country. The consortium saw the potential for their museums to play a critical role in the public's understanding of current research, if they could update their facilities and programs.

Progress had barely begun when the Bell Museum was hit by serious financial cuts. The situation came to a head in 1992, when the University's State Special appropriation that funded outreach and extension programs was cut. The museum's public programs were largely supported by this appropriation. More alarming, the College of Biological Sciences cut most of its funding for the museum as a whole. The cuts nearly spelled the end of the Bell's public programs. Birney and the advisory board responded by mounting an intensive letter-writing campaign. In the end, the State Special appropriation was restored, but the cuts from the college remained. Birney would have none of it, and in the fight with the college that ensued, he was forced from his position as museum director.

The advisory board and the James Ford Bell Foundation then stepped in to help. They formed a select committee

Don Gilbertson had a keen interest in developing educational programs for the public. University of Minnesota Archives.

Elmer Birney at the opening of the exhibition In the Realm of the Wild: The Art of Bruno Liljefors of Sweden, 1988.

When Scott Lanyon came to the museum in 1995, he faced challenges on both the academic and public sides of the museum.

made up of national museum experts to review the Bell's relationship to the University. They were charged to look at all options, including separating from the University. University president Nils Hasselmo became personally involved, and eventually the committee was able to negotiate for the hiring of a new external museum director. The committee also recommended removing the museum from the College of Biological Sciences. When Scott Lanyon, an ornithologist from the Field Museum in Chicago, was hired as Bell director in 1995, this was the situation he inherited. Shortly after he arrived, the Bell Museum was moved to the College of Natural Resources, a college with a long history of supporting public outreach. Its dean was a strong advocate for the museum.

To enhance the value of the Bell Museum to the University, Lanyon negotiated to have all new curator positions cross-appointed into appropriate departments in the College of Biological Sciences and College of Natural Resources. Lanyon also started an intensive strategic planning process with the objective of reimagining the museum. By 2001, following the recommendations of an architectural study and a series of focus groups, the Bell Museum, the University, and its community supporters came to the consensus that the museum could not achieve its mission at its old location. The vision was to create a new facility that would redefine the natural history museum for the twenty-first century.

By 2005, private fundraising was well under way, a site had been selected on the corner of Cleveland and Larpenteur Avenues on the University's St. Paul Campus, and architects were engaged to prepare preliminary designs. The University's new president, Robert Bruininks, was a strong supporter, and the new museum project began working its way up the University's list of priority capital projects.

With the architectural plans complete, plans for the new Bell Museum were submitted with the University's capital request to the Minnesota Legislature in 2008. The project received impassioned testimony from sponsors in the house and the senate, as well as a flood of letters and phone calls from museum supporters across the state. A supermajority of the legislature voted to appropriate funds for the project. It looked like a done deal, but then governor Tim Pawlenty used a line-item veto to cut the project's funding along with about sixty other initiatives.

Susan Weller welcomes visitors to a museum event.

David Hartwell, Ford Bell, Alice Hausman, and other supporters throw celebratory shovels of dirt during the Earth Day groundbreaking of the Bell Museum, April 22, 2016. Photograph by Kristal Leebrick, Park Bugle.

Lanyon, having done what he could and having served as Bell director for more than twelve years, decided to step down. Susan Weller, Bell curator of entomology, stepped in. The next year, the Bell Museum project was again approved by the legislature. Representative Alice Hausman, the museum's major sponsor in the legislature, and President Bruininks thought they had an agreement from Governor Pawlenty. But the governor vetoed the museum's appropriation a second time, stating that Minnesota did not need another Science Museum or Minnesota Zoo.

Failing twice at the capital is almost certain death for any University project. The museum being struck from the University's list of priority capital projects was an extreme low point for museum staff, its board, and its members. Many had worked for more than a decade to build a case for the new museum. Now Weller faced the difficult task of restoring staff morale and bringing the public back to the old Bell Museum after having been told for years that the place was falling apart and about to close. Staff had to revive exhibition and education programs and channel their energies into reexamining ideas for renovations at the old building, none of which adequately met the needs of a reimagined museum for the twenty-first century.

But other forces were at work that would turn the museum's fortunes around. The chair of the Minnesota Planetarium Society approached President Bruininks with the idea of a merger. Bruininks was a good friend of the museum and knew of its troubles. He also knew that the state had approved $22.5 million in bonding to build a new home for the planetarium on the roof of the new downtown Minneapolis Central Library but that the Planetarium Society had failed to secure matching funds.

Could these two struggling organizations be stronger and more effective together? Could the two—working as one—finally get the facility they both needed? The seed was planted. Bruininks asked Weller to explore the idea with the institutions' two boards. In fall 2011, the Minnesota Planetarium Society formally joined the Bell Museum.

The final breakthrough came in 2014 when Minnesotans elected a new governor, Mark Dayton, who was sympathetic to education and the environment. Seeing an opportunity to revive the civic initiatives the outgoing governor had vetoed, Representative Hausman invited all the projects that had been axed to reintroduce their proposals. The Bell Museum and Planetarium Society put together a new plan for

Museum advisory board chair Lee Pfannmuller and past chair Steve Birke at the groundbreaking ceremony. Photograph by Gary Seim.

One of the first parts of the new building to be constructed was the large concrete cylinder that would house the new planetarium. Courtesy of McGough Construction.

a facility that would house them both, and again approached the legislature. Hausman recalled the moment. "I remember," she said, "a visit to my office of representatives from the planetarium and the Bell Museum. They told me of their hopes and dreams for a joint building. From that point on, I was totally committed."

Over the next several months, with Representative Hausman applying pressure on her colleagues, and museum board members working the halls of the capitol, a funding deal was hammered out, passed by both chambers, and signed by the governor. Against what seemed like all odds, the Bell Museum, now combined with a planetarium, had prevailed.

In summer 2018, after nearly two decades of struggle, the Bell opened its new facility on the University's St. Paul campus. The staff who lived through and survived the whole ordeal now think that the governor's veto was a blessing in disguise: the new museum is far superior to the one planned in 2008. Perhaps most important, the museum is now seen as a major asset for the University, a gateway to the St. Paul campus, and a window to University research.

Encouragingly, the Bell is not the only university natural history museum to transform itself. New facilities have opened at the Universities of Oklahoma, Florida, Utah, Michigan, and Washington, and several others have undertaken major renovations. University-based natural history museums have taken renewed leadership in advancing discovery of the natural world and in bringing a better understanding of scientific research to the public.

Visitors line up at the front door of the new Bell Museum on opening day, July 18, 2018. Photograph by Joe Szurszewski.

From Earth to the Cosmos

The Journey of Minnesota's Planetarium

Call it a cosmic convergence. From the rock and mineral room on the fourth floor of the Minneapolis Central Library in 1950 to its featured place at the University of Minnesota's Bell Museum, the Minnesota Planetarium has traveled a rough road marked by ups and downs, dead ends, and near washouts. Its merging in 2011 with the Bell, an institution that has had its own ups and downs, was a long time coming. The convergence was the coming together of two beloved public institutions that had been marginalized by decades of tight budgets and political ill winds. The result has been a profoundly strengthened organization.

But in 1950, no one could have foreseen the events and ensuing struggles that were to come. When the Minneapolis Planetarium opened in what was billed as a science museum on the fourth floor of the old Minneapolis downtown central library, it was one of only fifty planetariums across the country. It seated just over fifty spectators, who viewed the heavens on a twenty-four-foot canvas dome that was lowered from a high ceiling for planetarium shows projected onto it by a Spitz A-1 projector. The projector's twelve-sided shape had been developed by inventor Armand Spitz (with input from Albert Einstein). Hundreds of tiny holes drilled into the star ball projected a *Stars Over Minneapolis* sky when lit in

The Minneapolis Planetarium at the Central Library, 1980s. Photograph by Rodney Nerdahl.

a darkened room. In its first year, more than four hundred people attended its shows.

The planetarium really took off when Maxine Haarstick was hired as a planetarium educator in 1951. Haarstick, a Twin Cities native and graduate of the University of Minnesota, had taught high school science in greater Minnesota before coming to the planetarium. Her innovative programming was groundbreaking—as she put it, "good basic astronomy that has a little bit of excitement."

This was a time of rapid growth for Minneapolis, and its central library was approaching capacity. The Soviet Union's launch in 1957 of Sputnik, the world's first satellite, had unleashed a flood of federal funding for science education, including support for planetariums. As planning began for a new downtown library, Haarstick was instrumental in making sure that it included adequate space for a planetarium. "Adequate space," in this case, was for a much bigger planetarium, seating 160 viewers under a forty-foot dome.

A projector did not exist that would work in a planetarium of this size. To fill the need, Spitz developed a custom model that projected over six thousand stars. Looking something like a giant mechanical insect, this was the projector that many visitors came to know and love as the "ant." It was suspended from the planetarium's dome, which had a cardboard silhouette of the Minneapolis skyline along its horizon to help orient audiences.

Opened in 1961, the new library had longer hours, and Haarstick expanded the show schedule to weekends. Attendance grew sixfold, from twenty-five thousand in 1959 (the last full year before construction) to almost one hundred fifty thousand in 1961. Programming continued to expand, with the themes in the planetarium's signature show, *Stars Over Minneapolis*, changing monthly to coincide with current topics in astronomy. Using federal Head Start funds, special shows were developed for inner-city students. Grants from Minneapolis Public Schools and NASA funded exhibits and shows for middle school and high school students. By the time Haarstick retired in 1979, she had become one of the best known and most beloved planetarium directors in the United States.

Maxine Haarstick with the custom Spitz projector, 1961. Courtesy of Hennepin County Library.

Rodney Nerdahl was hired to run planetarium operations in 1975. He produced and presented programs and cared for the planetarium's technology for the next twenty-seven years. Among the space topics presented during his tenure were Native American starlore, the connection of astronomy to the arts, dinosaurs, and the history of the Soviet space program. But Nerdahl felt that people came to the planetarium mainly to see the stars and ask questions, and this remained the heart of the program. During this time, the planetarium staff hosted events in state parks and other venues to view eclipses, meteor showers, and planetary conjunctions. In 1989, laser light shows were added to increase revenues and proved to be a fan favorite.

University astronomy professor Larry Rudnick at the July 2018 grand opening of the Bell Museum. Photograph by Joe Szurszewski.

STRUGGLES TO STAY ALIVE

All of this programming and technology delighted visitors—over two million had come to the planetarium over three decades. But behind the scenes it was a struggle to keep going. Between 1969 and 1972, the Library Board appealed to the state legislature for funding, but no action was taken. In 1974, the Library Board contracted with the Science Museum of Minnesota (SMM) in hopes of expanding and improving the planetarium programs. In 1980, the SMM renamed the facility the Children's Center and Planetarium. But the Science Museum was struggling with its own financial challenges, and in 1982 decided to cease operating the facility. The Library Board responded by putting together an emergency agreement to save the planetarium, which had strong support from then Minneapolis mayor Don Fraser and corporate donors. The volunteer group Friends of the Minneapolis Library agreed to assume responsibility for both operations and fundraising to keep the planetarium running.

The Friends then turned to a young University of Minnesota astronomy professor, Larry Rudnick, to help them navigate the new challenge they had taken on. Rudnick, a man with a passion for teaching his science, was instrumental in putting together an advisory committee and assumed leadership of the group. His relationship with the planetarium as a leader, advocate, and advisor has continued to the present, nearly forty years.

By century's end, the planetarium was falling behind in current technology. Shows were occasionally accompanied by sparks shooting from the old star projector. The Minneapolis Central Library was aging, too. By 2000, a plan and campaign were launched for a new library, with the possibility, but no guarantee, that a new planetarium would be included. Still, successive mayors and statewide stakeholders continued to support the need for a new planetarium. In 2002, the Minnesota Legislature approved $9.5 million in funding to build a new planetarium as part of the library. But then governor Tim Pawlenty line-item vetoed the planetarium's funding even as he approved the library's construction funds.

Deflated but not defeated, the planetarium's supporters regrouped and formed the Minnesota Planetarium Society (MNPS). Parke Kunkle, an astronomy instructor at Minneapolis Community and Technical College and another fierce advocate for a permanent planetarium, became president of the society's board of directors. Under Kunkle's leadership, a small staff was hired, and together with the board, they pushed to have the new library redesigned so that a planetarium could someday be added to the top of the building.

Sally Brummel and the ExploraDome at Parkview Center School in Roseville, Minnesota, 2016. Since initiated in 2006, the traveling dome has amazed more than two hundred thousand students, one classroom at a time. Courtesy of Roseville Public Schools.

Planetarium programs coordinator Sarah Komperud takes an audience on a journey through the Milky Way in the portable ExploraDome at the Minnesota State Fair, August 2017. Photograph by Patrick J. O'Leary.

In 2005, the Minnesota Legislature again approved funding for a new planetarium, this time for $22 million, but with the caveat that the society had to raise $19 million. Plans, however, had to be put on hold for a year as the Minneapolis Library system completed a merger with the Hennepin County Library system.

Then the MNPS was dealt a further blow when a feasibility study determined that a partnership with the library in its downtown location would not be financially sustainable. The society was forced to start looking for another partner, one that could share its mission to inspire, inform, and educate visitors about our planet, our universe, and the many benefits of scientific thinking, technological advancement, and exploration.

FINDING A HOME AT THE BELL

To keep the planetarium in the public's consciousness while searching for a permanent home, in 2006 the MNPS launched the ExploraDome program, directed by Joel Halvorson and Sally Brummel, both of whom worked for the society. The ExploraDome was the first of its kind: a digital portable planetarium that brought new technology to school gymnasiums. Students could cruise the cosmos without ever leaving their schools.

This program wowed students and teachers alike. Schools integrated ExploraDome visits into their astronomy curriculum and scheduled it to come back year after year. In this way, the staff and the MNPS board kept planetarium programming alive for more than twenty thousand people a year. The traveling program also had the added effect of continually teasing the public with grander possibilities. If stargazing was such a great experience under an eighteen-foot dome for just thirty people lying on their backs, how amazing would it be to experience that in theater seats under a dome three times that size? The ExploraDome served both to provide astronomy

The 120-seat Whitney and Elizabeth MacMillan Planetarium is a central feature of the new Bell Museum. Photograph by Joe Szurszewski.

The museum's roof deck is a great location for stargazing programs. Photograph by Joe Szurszewski.

lessons to students and to tempt the community with what *could* be.

In 2010, the planetarium finally found a promising match. Like the planetarium, the Bell Museum had had its funding for a new building erased by a governor's veto. But the stakeholders for both institutions had never abandoned their dreams. Under the leadership of University president Robert Bruininks, the two groups came together with a plan: by working together as a merged, larger institution, they could present a far stronger argument than either could acting alone.

But how would that work? The Bell Museum's focus had always been ecology, nature, wildlife, and environmental issues. How would astronomy and physics fit with that? In discussions between the two institutions, Halvorson pointed out that one of the most important outcomes of the space program was the ability to study Earth from orbit. Data from satellites have greatly expanded our knowledge of weather, climate, oceans, habitat loss, and animal migrations. A key value shared by the two institutions was a commitment to interpreter-led programming, where visitors could have personal experiences with a knowledgeable guide through nature or the cosmos.

In 2011, a deal was sealed, and the Bell Museum would continue to bring the ExploraDome to schools statewide. Not long after their merger, the museum installed a mini-planetarium using the same ExploraDome technology in one of its lower-floor classrooms. For the first time in ten years, Minneapolis again had a public planetarium, even if it was only a small one, accommodating only fifteen people at a time under a thirteen-foot dome.

Finally, in 2014, with new governor Mark Dayton committed to civic support of science education and the environment, and backed by a sympathetic legislature, the merged institutions got their breakthrough. Plans for a new building that would house them both were assembled, and a yearlong effort was launched to secure the funding. In 2018, a new Bell Museum and planetarium opened its doors to the world.

Today, the new Bell Museum houses the Whitney and Elizabeth MacMillan Planetarium, with 120 seats under a 52.5-foot dome. A key architectural feature of the museum, the new planetarium presents digital projections ranging from exploring galaxies billions of light-years away to examining single cells inside the human body. Anything that can be modeled in three dimensions, from the surface of Pluto to a forest in the north woods, can be explained and viewed using the planetarium's technology. Sitting in a planetarium, with the image filling your visual field, you feel that you are no longer a remote observer but actually there and part of the action.

The first planetarium show in the new museum was *Minnesota in the Cosmos*, a feature presentation that tells the cosmological and geological story of Minnesota. In the museum's first year of operation, planetarium staff created over a dozen original productions enjoyed by over one hundred thousand visitors. All programs included live facilitated presentations, keeping with the storytelling legacy of Rodney Nerdahl and Maxine Haarstick.

The planetarium is truly a testament to the remarkable perseverance and sustained commitment of its friends and beloved fans. Generations of passionate supporters have worked tirelessly to keep the institution alive—they never stopped believing. The goal has remained the same: to share the wonder of the stars and connect the people of Minnesota to their universe.

THE RIDE OF HIS LIFE

On February 4, 1961, George "Pinky" Nelson climbed into the family car to join his parents on the ninety-three-mile trip to Minneapolis to attend the opening day of the Minneapolis Planetarium. Only ten at the time, the future NASA astronaut had no way of knowing how epic that journey would be for him. The country was deep into the Cold War, and the Soviet Union had already launched Sputnik, the first artificial satellite into orbit around Earth. The space race was on, and the nation was determined to catch up. The show young Nelson experienced that day under the planetarium's darkened twenty-four-foot canvas dome electrified him, firing his imagination. "That Journey to the Stars [program]," he later recalled, "had a profound impact on me, sparking a lifelong interest in space and a career as an astronomer and NASA astronaut. Every child should have the chance to experience the same feelings of awe and curiosity I felt that day."

George "Pinky" Nelson was inspired to become an astronaut after attending the Minneapolis Planetarium when he was ten years old in 1961. Copyright by The NASA Library/Alamy.

The Road to a Reimagined Museum

... ignite curiosity and wonder, explore our connections to nature and the universe, and create a better future for our evolving world.
—Bell Museum mission statement, 2017

In the 1990s, the Bell Museum, like many natural history museums across the country, was facing big challenges. Society was becoming more urban, institutions were working to be more inclusive of diverse communities, and technology was changing how people learned. Many children were growing up largely disconnected from the natural world.

For most of a century, natural history museums had filled their darkened halls with dioramas depicting pristine nature, or dinosaur halls presenting concepts of classification and evolution. Were these galleries still relevant in a world threatened by pollution, habitat loss, resource exploitation, population growth, biodiversity loss, and climate change? How should museums, which had long been dedicated to exploring, documenting and displaying Earth's diversity, respond to forces that were threatening the basic ecosystems of the planet?

When Scott Lanyon became museum director in 1995, he was familiar with these trends from his experience at the Field Museum of Natural History in Chicago. He knew also that no museum had come up with the perfect solution. Lanyon could see that the Bell Museum needed major change and probably a new home. But it was going to take time to build support for this among museum staff, the University, and the broader community. Hence, he started what would eventually become twenty years of planning, evaluation, outside consultation, and experimentation. This effort would finally bear fruit in a new building on the University's St. Paul campus, with a re-envisioned exhibition gallery.

One of Lanyon's first moves was to commission an audience study. The results were a wake-up call for museum staff. Most first-time visitors described the museum as dark, old, tired, and not inviting. Despite the museum's record of

The darkened halls of the old Bell Museum created a contemplative atmosphere for visitors.

The bird observation station was one of a number of experimental projects completed in the 1990s that added interactive elements to the diorama halls.

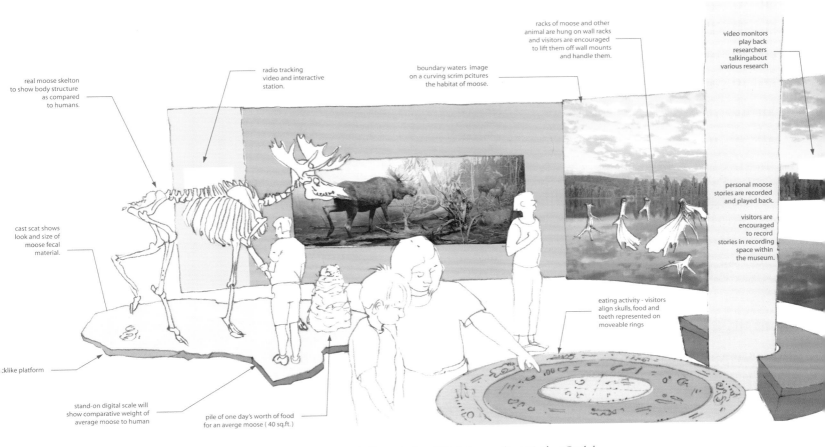

One of many sketches that explored how the dioramas might be re-envisioned. Illustration by exhibit design consultant Matthew Groshek.

success with temporary and traveling exhibits and innovative education programs, the public's perception of the Bell Museum was still largely defined by its traditional diorama halls. The future of the museum would hinge on finding ways to reimagine the long-neglected permanent galleries. Should the museum dispense with the dioramas? Should it be more like a science center, a children's museum, or maybe a nature center? Some even suggested the museum give up on a physical home and just focus on its outreach programs. The staff wrestled with these questions.

In 1998, Lanyon arranged for a major external evaluation of the museum's public programs. The study found the museum in deep trouble and struggling to meet its mission. The evaluators pointed to the need to define a clear and unified vision. The museum's connection to the University was seen as a key opportunity that was not fully utilized. The study also noted that the role for dioramas in the future of the museum needed to be resolved.

In 2000, Lanyon convened another panel, this one of national museum leaders, to meet with staff, museum board members, donors, and University faculty and administrators. They identified similar issues and helped to point toward potential solutions. This was the first major effort to engage key leaders in the community. Sifting out of the long list of findings and recommendations were several key challenges the museum needed to meet: define the museum's role in addressing environmental issues, find ways to make the dioramas more engaging to modern audiences, and identify more effective ways to

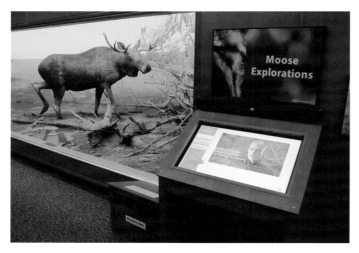 

A prototype of a touch-screen interactive paired with a diorama was tested in the old museum.

The exhibition Exploring Nature's Histories and Mysteries *demonstrated the value of direct interaction between visitors and University scientists. Graduate student Jacob Egge speaks with a visitor while working in the exhibit's "research shed."*

present current science research to the public by building on the museum's connection to the University.

The museum already had some experience with presenting controversial environmental and social issues, such as invasive species and AIDS, in its temporary exhibitions. Among the key lessons learned from these exhibits were to use humor whenever possible, to provide role models for visitors, and to indicate clear action steps. But museum staff still argued over the value of the dioramas, although there were little objective data to back up anyone's opinions.

With this in mind, Don Luce, the museum's curator of exhibits, used a professional development leave to review the latest findings from museum research and to explore innovative ways for using dioramas to present environmental issues. Museums were just beginning to experiment with renovations of their diorama halls. Luce studied their evaluation reports and visited museums in North America and Europe to observe their exhibitions and interview key staff. Visitor research was beginning to show that dioramas, when integrated with other types of experiences, were effective learning tools. He also saw a variety of innovative ways to use natural history collections to inspire and engage visitors.

Knowing that the dioramas were one of the most distinctive and extraordinary assets of the Bell Museum, Barbara Coffin, head of adult programs, and Luce decided to test ideas for reimagining the dioramas. They selected the moose diorama for reinterpretation. After the redesign, visitors could lift and feel the weight of moose antlers and try to identify small elements from the diorama scene by touch. A key component of the new design was a touch screen with a suite of programs that interpreted the diorama in a variety of ways. It included several search activities designed to engage families and an embedded series of short videos that followed a current research project. An evaluation compared visitor behavior at the wolves diorama, which lacked any additions, to the newly augmented moose diorama. The results were dramatic. Visitors spent three times as much time at the moose diorama. They used the hands-on elements and the touch screen but also spent more time observing the diorama. This study proved to skeptics that the dioramas could be effective learning tools for presenting science. It also suggested that in a new building the dioramas could be rearranged and integrated with a variety of exhibits to create exciting new visitor experiences.

But could the Bell Museum, with access to a host of

The exhibition Exploring Nature's Histories and Mysteries *revealed the museum's often hidden world of collections and research.*

Using life-sized photographs of researchers, field equipment, and actual lion specimens, The Lion's Mane *demonstrated a successful approach for bringing University science to the public.*

University scientists and museum curators, do more to improve the public's understanding of current research? Could the museum show visitors "science in action"? Many of the museum's external consultants were skeptical. To find out, the museum planned a large, new, temporary exhibition, *Exploring Nature's Histories and Mysteries*. The objective was to feature samples from all the museum's collections, from birds to butterflies, mussels to mammals, and flowers to fungi. At the center of the gallery was a re-created lab where researchers could work while also interacting with visitors. During the six-month run of the show, each curator committed to having a presence in the gallery for several weeks. Evaluation of the exhibition found that visitors thought their personal contact with a scientist was a valuable and unique experience. But curators and staff questioned whether maintaining this kind of experience on a long-term, ongoing basis was practical.

Next the museum experimented with alternative ways to present current research. *The Lion's Mane* exhibition engaged visitors in the process of field research by following the steps of University ecologist Craig Packer and graduate student Payton West as they pursued what seemed like a simple question: why do lions have manes? Another exhibition, *Mysteries in the Mud*, followed researcher Bryan Shuman, who was using cores from lake sediments to find clues to past climate change. One of the exhibit's most successful components was a small video installation that followed the researcher from his initial interest in nature as a young adult through his career as a research scientist. These exhibitions were evaluated to assess how well they achieved their goals. Successful strategies were to focus on the researcher, rather than the research, to communicate the scientist's passion for nature and discovery, and to engage visitors in simple, hands-on steps that demonstrate the practice of science.

By the time funding for a new building was finally secured in 2014, the Minnesota Planetarium Society had merged with the museum, dramatically broadening the scope of the museum's public programs. Initial plans called for an "Earth and Space" gallery to present astronomy, geology, and related physical sciences separate from the natural history exhibits. But this seemed disjointed. The planning team wondered if there might be a way to better integrate exhibits of the now combined institutions. They held a series of meetings to formulate a unified description of a vision for the new museum's visitor experience. Here the concept of a journey first

Based on experimentation in the old museum, an interactive touch screen with content for all ages is part of each diorama in the new Bell. Photograph by Joe Szurszewski.

started to take shape. When Gallagher & Associates were hired as exhibit designers for the project, they also urged the museum to work to create a single, coherent storyline. By using a timeline to organize the content, the new museum's exhibits could be presented within a series of linked galleries, with a single storyline—a storyline that would take visitors on a journey, a journey of discovery through space and time, from the beginnings of the universe to the future. It was a compelling vision and was also a distinctly different approach from other museums in the Twin Cities area.

All the years of planning have paid off. As visitors explore the Minnesota Journeys galleries in the new museum, they experience the results from decades of testing and experimentation. The display of biological specimens of the museum's scientific collections in the Diversity Wall can be traced back to the collection cases in the *Histories and Mysteries* exhibition. The Discovery Stations featuring University researchers were first tried out and tested in the *Mysteries in the Mud* exhibit. The use of small dioramas as walk-around, immersive displays, prototyped in the *Peregrine Falcon* exhibition, can now be seen throughout the Web of Life gallery. Each large diorama includes environmental stories, hands-on objects, and an interactive touch-screen interpretation. Dioramas are no longer presented as rows of windows in a darkened hallway but as vista points on a tour of the state.

A Minnesota thread runs throughout the gallery, from the 1.8 billion-year-old stromatolites from the Iron Range to the current biodiversity experiment at the University's Cedar Creek Ecosystem Science Reserve. The last section of the Minnesota Journeys galleries, Imagine the Future, is designed to be flexible and easy to update. Here, visitors can apply what they have learned about how nature works to imagine a sustainable future. They see University scientists and ordinary citizens working to find solutions to present-day environmental challenges. Activity carts throughout the galleries give visitors a taste of the science process, and at periodic special events, museum and University researchers bring their collections, tools, and passion for discovery to the public.

Gayfeather flowers (Liatris) at Touch the Sky prairie near Luverne, Minnesota, by Jim Brandenburg. Almost all the photographs and much of the video footage in the museum's Minnesota Journeys galleries are drawn from Brandenburg's fifty-plus year career exploring and capturing the beauty of Minnesota's wildlife and wild places.

The design of the building includes two large "story box" windows that project over the first floor. They invite views into the exhibits from the outside. In this case, the woolly mammoth diorama is seen from across the street. Photograph by James Steinkamp Photography.

Designing with Nature

The Bell Museum's New Home

There's a new museum in the Twin Cities, and it's a beauty. . . . the sleek, sustainable exterior is a mix of gleaming bird-safe windows, rustic cooked white pine from Cass Lake and weathered steel from the Iron Range.
—MinnPost, *June 27, 2018*

Can buildings make statements? Judging by the new, state-of-the-art Bell Museum on the corner of Larpenteur and Cleveland Avenues in St. Paul, the answer is an emphatic "yes!"

The Bell Museum opened the doors of its new facility on July 18, 2018—its fifth home over the course of its nearly 150-year existence on the University of Minnesota Twin Cities campus. Together the building and the surrounding site demonstrate how the design of a built environment can coexist sustainably with both the natural and cultural landscape. The building design not only incorporates energy standards of the twenty-first century and a commitment to locally sourced building materials, it also honors important legacies of this beloved museum.

Perkins and Will, a well-known architectural firm, designed the project, which includes features that echo the archetypal style of the museum's magnificent Francis Lee Jaques habitat dioramas. As David Dimond, the principal design architect, said, "We wanted the building to be a series of 'story boxes' that would captivate audiences on the outside just as Jaques's dioramas do on the inside."

Indeed, those story boxes are an integral part of the building's design, connecting the outside to the inside, and the inside to the outside. The building's north-facing story box contains a life-sized woolly mammoth, along with other Ice Age megafauna, set before a receding glacier. It is a diorama that visitors can literally walk into. The ecological story depicted in this scene can be viewed from nearby streets, especially at night when illuminated. In this way, the building becomes an important part of the story the museum was created to tell: Minnesota's natural history as a journey through time and space.

One of the first things visitors may notice is the white pine and iron siding on the exterior of the museum. Part of the design team's goal was to create an environmentally sustainable building, using locally sourced materials whenever possible. Forty percent of the museum's exterior is covered with thermally modified Minnesota white pine. The process of turning a soft wood such as white pine into a durable long-lasting siding involves cooking the wood in a giant kiln. This chemical-free process drives out the naturally occurring sugars, making the wood resistant to both decay and weather. The siding was produced by a company located in the northern Minnesota community of Palisade, with wood sourced from ecologically managed Minnesota forests certified by the Forest Stewardship Council. The remainder of the exterior siding is likewise reflective of Minnesota history and its legacy of ore mining. The steel siding made from iron ore has weathered naturally to create a permanent, protective, rust-colored patina.

But using locally sourced materials was only part of the design team's bigger commitment to creating an energy-efficient design for this twenty-first-century building. The new museum is what is known as a "B3" building. Leading the nation, Minnesota requires all building projects that receive state funding to adhere to a progressive energy standard designed to significantly reduce their energy and carbon footprints. The goal is to make all state-constructed buildings carbon neutral by 2030. Buildings built in the years 2016–18, such as the Bell Museum, were required to be 70 percent more energy efficient than the average state building of that time.

The project reduced the building's fossil fuel use by more than the minimum specified in the energy efficiency requirements. A combination of design features and building materials were used, including energy-efficient windows, 100 percent LED lighting, and energy-efficient audiovisual equipment. A cutting-edge configuration of heating and cooling elements coupled with a dedicated outside air system and roof solar panels contribute to the building's renewable energy features.

The efficient use and reuse of water were important design elements of the sustainability goals. Clean rainwater that falls on the building's roof is channeled to a naturalized pond that supports a diversity of aquatic species and provides irrigation water for the site's native and adapted plantings. Stormwater that drains off the parking lot is directed into rain gardens or permeable pavement, where the contaminated water is filtered before replenishing groundwater.

From the beginning of the project, the Bell Museum design team acknowledged the critical importance of the interplay between natural and engineered systems. "Whether we want to acknowledge it or not, invasive human activity and climate change are irreversibly damaging our Earth and its diverse ecosystems," said Doug Pierce, architect and resilient design director with Perkins and Will. "The Bell Museum recognizes this and is demonstrating pioneering leadership in the area of resilient planning and design in an effort to educate around, advocate for, and advance healthy and viable ecosystems." The use of bird-safe glass is an example of this leadership. A custom visual frit pattern was developed that has effectively deterred bird strikes. Frit patterns also minimize heat transfer to the inside on sunny days, thus having the added benefit of improving energy efficiency.

Deliberate planning also guided the design of a learning landscape on the five acres surrounding the museum. Only plants native to Minnesota were used in the plantings. Seed mixes of native prairie species suitable for full sun or shade in either wet or dry habitats were specially formulated for pollinator gardens and rain gardens. A demonstration living roof is located on the observation deck. Living, or green, roofs increase pollinator habitat, reduce roof temperature, conserve energy, reduce stormwater runoff, and improve filtration.

Important parts of the state's early geologic history are on display in the garden plazas north and south of the glass-walled lobby. In the Big Woods garden to the south sit huge cylinders of Morton Gneiss, a 3.5-billion-year-old granite that is the oldest rock exposure in North America, and Platteville Limestone, a sedimentary rock formed 450 million years ago when an ocean covered Minnesota. Large boulders,

Heat-treated white pine, sustainably harvested in Minnesota, resists weather and decay without chemical treatment. Courtesy of Perkins and Will.

Unseen by visitors are the solar panels on the museum's roof that contribute to the building's electric energy use. Courtesy of Perkins and Will.

Front entrance of the Bell Museum, looking east. Special glass and window shades reduce heat gain. Photograph copyright Peter J. Sieger.

called glacial erratics, that were excavated from the museum site during construction are also a part of the outdoor learning landscape. Glacial erratics are a much more recent part of Minnesota's geologic history. They were left on the Minnesota landscape 11,000 years ago when the glaciers retreated.

In the north-woods garden on the north side of the building, a walk-through outside diorama depicts a predator–prey encounter between two of Minnesota's iconic wildlife species, the moose and the wolf. These outdoor sculptures, created by the museum's former exhibits preparator Ian Dudley, were favorites at the museum's Minneapolis location. At the new site, three raven sculptures were added, expanding the ecological story of this classic north-woods scene.

Minnesota's nature is woven into the museum's design inside and out. Bell Museum science director George Weiblen says with satisfaction, "The Bell Museum building and the site's learning landscape demonstrate how we can learn from nature . . . to thrive with diversity, live within limits, and adapt to change. This is our promise for the future." The Bell Museum's new home stands ready to serve Minnesotans for the next hundred years as society explores the challenges of a changing environment.

North-woods plaza with moose, wolves, and raven sculpture. Photograph by James Steinkamp Photography.

Patterns on the window glass, including dots and horizontal lines, help prevent bird collisions. Courtesy of Perkins and Will.

Native landscaping, including a pond, provides wildlife habitat and filters stormwater. Courtesy of Perkins and Will.

The lobby opens out to plazas with native plantings to the north and south. Photograph by James Steinkamp Photography.

Moose being craned into the new Bell Museum, August 2017.

Moving Minnesota

Dioramas in a New Habitat

It has been said that leaving home is never easy. For the Bell Museum that was particularly true. Indeed, "never easy" hardly captures what the museum went through in its move to St. Paul.

Concern had been growing for some time that to serve contemporary audiences and those into the future, the Bell Museum would have to leave its longtime home on the Minneapolis campus for more flexible and modern quarters. Moving the museum brought up an important question. Could the Bell Museum's renowned dioramas be moved? Built to be permanently in place, never to be moved, how in the world could they ever be relocated? Was it even possible? By the late 1990s, as plans for a new facility on the University's St. Paul campus were being considered, museum staff and the administration took a collective breath and dove into an assessment of a diorama move.

The first step was to gather as much information as possible about how the dioramas were made. Some of the best information came from a series of slides taken in the 1940s during the construction of the maple–basswood forest (now called Big Woods) diorama. The images showed that the foreground was constructed on a lightweight framework of wooden boards. Large objects, such as trees and large animal mounts, were set in place first and then the framework was built around them. The framework rested directly on

This photograph of John Jarosz from 1949 provided information about how the diorama foregrounds were constructed.

the concrete floor and was not tied together in a way that would make it possible to move the foreground as a unit. Some of the background murals were adhered with hide glue, which was workable, but others, including most, if not all, of the murals painted by famed diorama artist Francis Lee Jaques, were adhered using a lead paste adhesive. This was much more permanent but also toxic to remove—not an encouraging picture.

The next step was to see if any other institutions had moved dioramas. Don Luce, curator of exhibits, searched for other museums that had taken on similar projects. Luce visited the Canadian Museum of Nature in Ottawa, which was in the midst of moving its dioramas, but these dioramas were constructed quite differently from the Bell's. He learned that the Illinois State Museum had moved dioramas that were constructed similarly to those at the Bell Museum. At both museums the diorama walls had been constructed on steel upright supports with metal lathe and concrete. This was overlaid with plaster, and finally the canvas was glued to the plaster. This type of construction made for strong but very heavy walls. If the walls were to inadvertently flex, it would crack the plaster, and the murals would be ruined. All in all, the fact that dioramas had been moved at other museums helped to relieve some of the mounting anxiety. But none of these dioramas was exactly like the Bell's, and none had been moved across town to a new building.

Terry Brown disassembles the branches of the basswood tree in the Big Woods diorama, 2017.

The deer fawn, birds, and wax plants were removed from the diorama foregrounds, numbered, and placed on trays.

As planning and fundraising for the new Bell Museum picked up, a team of experts was assembled, and the museum was closed for a few days while the diorama windows were removed to learn as much as possible. It became clear the murals could not be safely separated from the wall. The walls and the murals would need to be moved together. To remove them from the building, the walls would need to be cut. Toxins were found, but fortunately their concentrations were low.

A plan for how to move the dioramas was worked out, but many questions still remained that could not be answered until the diorama foregrounds were disassembled and the background murals were moved. A test move of one or two dioramas would be required. In early 2016, the museum was temporarily closed, and the front walls of the Dall Sheep and Beaver dioramas were removed. The Dall Sheep diorama was never planned to be moved, so it became the "sacrificial lamb." It would be used to test the methods for moving the background mural. The Beaver diorama, built in 1917, was used to test the protocols for handling the hundred-year-old wax plant models, which had become brittle with age.

The test move was also a chance to test the team of experts who would do the work. Trevor Dickie was the University's project manager for the construction of the new museum, and McGough Construction was the general contractor for the new building, including the diorama move. Tom Amble, exhibit preparator for the museum, would be the on-site manager for the move. Split Rock Studios, a local exhibits company with a history of building dioramas, would manage the foreground move with Terry Chase, from Chase Studios, and Terry Brown, from Museum Professionals. They had both spent much of their careers building dioramas. Luke Boehnke, and his team called Wolf Magritte, together with the art logistics firm Methods and Materials, would be in charge of moving the background murals. Boehnke had moved dioramas for museums in Canada. He would design and build the steel framework to hold and support the massive background walls, which weighed six thousand to eight thousand pounds. Kristy Jeffcoat and Megan Emery from Midwest Art Conservation Center (MACC) would supervise a large team of conservators, who would lovingly protect, clean, and restore the murals. All of these professionals saw the new Bell Museum as a once-in-a-lifetime project and worked with a passion for its success.

Being able to walk into the dioramas and see the murals as Jaques would have was a revelation to many of those working on the project.

To safely move the curved walls, a rigid steel framework, or armature, was constructed to support the massive structures.

THE MOVE BEGINS

In January 2017, the old museum closed its doors for good, and the race was on to disassemble, move, restore, and reassemble the dioramas and integrate them with new exhibitions for the museum's grand opening in summer 2018. The first step was for Chase and Brown and their assistants to meticulously remove the wax plants, small animal mounts, and other foreground elements. Each item was assigned a number, and its location marked on large photos. Once the small delicate items were removed, the walls around the front window of each diorama were demolished, revealing for the first time the underlying structure of the foreground.

In some cases, such as the Wolves and Snow Geese dioramas, the "ground form" was built in sections that could be taken apart. But in most cases, the ground form needed to be sawed into sections. Safety was a concern. As each new area or material was exposed, samples were taken and tested for hazardous chemicals. At first, team members wore hazmat suits with respirators. Dust was carefully contained, and large air filters were run continually to protect both the workers and the diorama elements. Luckily, the tests showed that hazard levels were low, and the crew could shed the hazmat suits and respirators for most of the work.

Once the foregrounds were cleared, work on the backgrounds began. The walls were twelve feet high, ten feet deep, and either twenty or twenty-two feet wide. Being inside revealed details of the masterful painting that were not visible from outside the glass. The crew from MACC cleaned every inch of the surface and then adhered a layer of tissue that would protect the painting surface from dust and abrasion.

In order not to lose any of the painting during cutting, two-inch-wide strips of the canvas were removed at the cut lines. A steel framework was welded together to support the large top section of the curved background wall. Once the framework was secure, the main section of the wall was raised with jacks about an inch until it hit the bottom of the floor above. The bottom of the diorama wall had to be cut at the center so it could be removed in two sections, one from each side. Once the bottom sections were removed, wheels were added to the main section, and it was lowered to the floor.

At any one time, most of these steps would be happening somewhere in the building. Chase and Brown might be involved in the delicate process of dismantling a diorama's foreground while walls were being cut and lowered elsewhere.

Each diorama background section was craned onto a flatbed truck and taken to the new building.

In the new building, the three wall sections were carefully realigned and then welded together.

Art conservator Kristy Jeffcoat touches up the cut lines on the Sandhill Cranes diorama, 2018.

Conservators Jeffcoat and Emery might be applying tissue liner to a mural while around the corner Boehnke and his crew would be sawing, grinding, and welding steel angle iron. Transparent dust barriers were put up and moved as needed, and the museum galleries were filled with stacked trays of wax plants and animal mounts sealed in plastic wrap.

The feathers and hides of many of the birds and mammals had faded over the years, so restoration specialist Brown set up a lab in the Touch and See Room, where he cleaned and recolored the animal mounts. Everyone was working toward the common goal of moving the dioramas to the new museum beginning in June 2017.

Large sections of the exterior of the old museum were removed on both the first and second floors. Finally, on June 20, 2017, the first of the diorama backgrounds was wheeled out of the opening, craned onto a waiting flatbed truck, and transported across town to the new museum on the St. Paul campus, where a section of the exterior wall on the second floor had been left unfinished. After the background murals, the large animal mounts, each sealed in plastic drapes, went out the side of the building. It was an amazing sight to see moose, elk, and wolves "fly" out of the old Bell Museum.

PUTTING IT ALL BACK TOGETHER

Once in the new building, the three sections of each background needed to be exactly realigned and then welded back together with new steel supports. The walls were then pushed into place within new enclosures. The cuts were filled and smoothed to make a flat surface. At this point, the painting conservators returned and glued the two-inch canvas strips back in place. They then filled and in-painted the cuts until the seams were invisible.

Next came the task of putting the jigsaw puzzle pieces of the foregrounds back together. These were assembled on new bases, so that the entire foreground could be moved as a unit, if needed, in the future. Some of the most difficult restorations were of the Moose and Tundra Swans diorama foregrounds. Both included muddy lakeshores, but the "mud" in the diorama, although beautifully formed and painted to look wet and soft, was made from brittle plaster. Wherever possible, cuts were made in the plaster in places that would not be visible. But many jagged and cracked edges needed to be repaired until they completely disappeared. Trees were set in place, then sections of the trunks were bolted together. Branches and twigs with delicate wax leaves were added to the trunks. Then each wax

Shannon Larson and Terry Brown clean the water surface of the Snow Geese diorama.

Terry Chase prepares wildflowers to be reinstalled in the Big Woods diorama.

wildflower, newly cleaned and restored, was glued into place.

Important improvements were made to the foregrounds. At the old museum, loose sand and dirt were used in many of the dioramas. This was now all glued in place so that it would not be a source of dust over time. New techniques were used to preserve mosses and leaves. In the Cascade River diorama, the trunk of the cedar tree had ended about a foot below the ceiling. Terry Brown made molds of the bark and extended the trunk, with more branches, to the ceiling.

One of the biggest improvements made was in the way water was simulated. In the Beaver diorama, first created in 1917, then moved into the old Bell in 1940, water was simulated with a dark-brown mixture of gelatin and glycerin. This gummy material had cracked and lifted up in places over time. In removing it, the team discovered that it was not original to the exhibit but must have been added in 1940. It was decided to make a new water surface for the beaver pond. Painted to match the background and then coated with a layer of acrylic polymer resin, the new pond "water" now included ripples from the swimming beaver.

The dioramas in the new museum are protected within sealed enclosures. The lighting has been replaced with modern LED lamps to bring out the beautiful colors of the murals and to highlight the foreground plants and animals. The front glass has been replaced with specially produced nonreflective glass that is actually more transparent than the old plate glass. Many visitors who knew the dioramas at the Church Street location have a hard time believing they are the same "old" dioramas.

During all steps of the move, the dioramas were treated as irreplaceable works of art and important cultural artifacts. The team worked hard to preserve and restore them in keeping with the intent of their original creators. Some critics of the project had argued that despite the restoration team's best efforts, the dioramas would never be the same. In a way they were right: the dioramas are not the same. Many would now argue that they are actually better.

A redstart from the Big Woods diorama before and after restoration. Photograph by Terry Brown.

A Journey

The story of the universe is, in many ways, the story of you. You are an inextricable part of the drama that is played out every minute of every day.

Your story began 13.8 billion years ago with a big bang. As our modest-sized Earth was formed billions of years later, it was bestowed with special characteristics that eventually gave birth to you. Water. Sunlight. Elements. Although our human curiosity drives us to search for life on other planets, so far we are alone in our watery world.

You really are unique.

Our planet itself is rich and diverse with species. Billions of years ago life arose from simple cells in a primordial free-for-all. Microorganisms photosynthesized, the atmosphere became breathable, and voila!—the stage was set for an explosion of species that today includes bacteria stewing along deep-ocean vents and lichen slowly colonizing alpine rocks.

Nature never stands still. Thousands of years ago, on this spot where you now stand, ice flowed, retreated, and ground down the earth. Catastrophic glacial events changed today's Minnesota, giving us cool forests, treeless prairies, and hearty rivers.

As you go about your day on this rich and robust Earth, you choose survival. Humans, just like herons, honeybees, or hares, must feed, reproduce, and raise their young. In our drive to survive, we consume and grow. Sometimes this growth is benign to systems and other species, sometimes not.

We as a species have triggered a new epoch in Earth's history, where humans are no longer mere inhabitants, but in many ways, drivers of change.

Every day you spin with the Earth. Every day you write your story. What will it be? Where will it end? How will it change the stories of others?

Mary Kroll, consultant to the exhibit team, composed a narrative that personalized the experience. It was read at the groundbreaking ceremony for the new Bell Museum on April 22, 2016.

The Experience

A Journey through Time

The designers of the Bell Museum's main exhibition space, the Minnesota Journeys galleries, had set themselves quite a challenge: how do you combine astronomy, geology, biology, and ecology all within a single story grounded in a place called Minnesota?

The concept they finally came up with was a journey that would take museum visitors on a trip through space and time. The journey would include the origin of the universe, formation of Earth, the rise of life, its evolution and diversification, and how all that came to express itself in the landscapes and ecosystems of Minnesota. It would end with a challenge to visitors to imagine the future.

Near the entrance to the gallery is a photograph taken by a powerful telescope. The telescope was focused on an empty spot between the stars in the sky and held there to gather light. It shows hundreds of smudges and swirls. Each one is a galaxy, each containing billions of stars. From this unimaginable vastness, we zero in on one particular speck of dust, our planet, Earth, and start our journey.

Our tour starts with a walk up the grand staircase.

As we enter the Life in the Universe gallery, the ceiling opens to a ribbon of stars. A model of Earth hangs overhead. Here we can explore the special features that allow Earth to be a home for life. Photograph by Andy Hardman.

There are millions of galaxies in the universe, each containing billions of stars. From this unimaginable vastness, we focus on one particular speck of dust, our planet, Earth. Photograph by Andy Hardman.

But there is more! Explore how our atmosphere captures infrared light, keeping our planet at a livable temperature. Nearby, see how volcanoes and mountain building bring to Earth's surface new minerals and elements, essential for life. Photograph by Joe Szurszewski.

A large "bowl" greets us as we enter the next section of the gallery. In it, the key features of our planet—liquid water, minerals, and energy—interact. These conditions favored the formation of complex organic molecules, the building blocks of life. Photograph by Joe Szurszewski.

Natural history collections are a catalog of diversity and key to understanding evolution. The wall of specimens in the Tree of Life gallery displays plants, skulls, insects, fishes, and fungi, a small sampling from the museum's scientific collections. Look, compare, and wonder. All of these species are related to each other through the great tree of life. Photograph by Andy Hardman.

Look into the eye of a life-size woolly mammoth, and imagine what it would have been like to be in Minnesota some twelve thousand years ago. Thick glaciers covered much of the state, and giant mammals roamed around the glacier edge. Photograph by Andy Hardman.

Walk through the big archway and into the northern forest in the Web of Life gallery. Explore how the interactions of plants, animals, and the environment form the web of life. This journey through Minnesota's three major biomes is an exploration of life's biological interconnections.

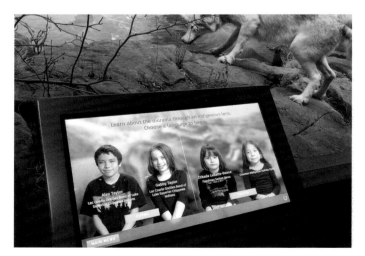

The dioramas are like vista points on the journey. Visitors can explore the details of the scenes using touch screens. Several include interpretations in Dakota and Ojibwe languages by local students.

Beyond the lynx and hare are the beavers. They are placid plant eaters but leave a big footprint on the forest. By cutting deciduous trees like aspen and birch and by building dams, beavers change the character of the forest and create habitat for many other species.

Just west of Lake Itasca, Minnesota's landscape changes dramatically. We are now at the edge of the prairie, and a bull elk is bugling a challenge to a rival male. Each year animals switch their attention from simply making a living to finding a mate and raising a family. Photograph by Andy Hardman.

The drive to find a mate has led to many amazing adaptations, such as the diversity of horns and antlers. Photograph by Andy Hardman.

Don't just look at the cranes doing their mating dance—see if you can learn the steps! Then turn around, and look into the nests of marsh birds to see their eggs and young.

Plants also need to find mates to reproduce. This case displays a small sample of the vast diversity of plant flowers and seeds.

Look into a patch of woods near Waseca, Minnesota. It is early spring: the tree leaves are just emerging, but the small plants on the forest floor are in full bloom. They need to gather light and set seed as soon as they can, before the trees leaf out and shade them. Look for warblers and other songbirds in the trees. Many have just arrived from the tropics. Photograph by Andy Hardman.

The Passenger Pigeon diorama marks a transition point in the journey and offers a poignant lesson from the past. How did we drive the most populous bird in America to extinction in just a few decades?

As we walk toward the Imagine the Future gallery, we are immersed in images that record humanity's exploding imprint on Earth. At night, lights from cities worldwide form a glowing web across the planet. Photograph by Joe Szurszewski.

In the last part of the journey, we are asked to consider three key ideas when imagining a sustainable future for our planet: living within limits, adapting to change, and thriving with diversity. Here we meet scientists, students, and citizens working to find solutions to society's grand environmental challenges. Photograph by Joe Szurszewski.

Afterword

Denise Young, Bell Museum Executive Director

The intersection of science and public curiosity is a powerful place.

For 150 years, the Bell Museum has played an important role in Minnesota, studying and interpreting the state's natural history. The museum's influence over time has spread to the wider world, as the stories in this book have documented in elegant detail. I have every confidence that it will continue to serve our state and beyond in vital and exciting ways for decades to come.

The stunning new Bell Museum facility that opened in summer 2018 is a fitting match for our mission to ignite curiosity and wonder, to explore our connections to nature and the universe, and to help create a better future for our evolving world. With two hundred thousand people visiting annually, the museum's platform is larger than ever—at a time when it could not be more relevant. Our global challenges are complex and urgent. From climate change to the world's increasing human population to treating diseases of pandemic scale, nature and science remain central to the solutions needed for our planet. The Bell Museum remains a positive force in this unfolding story, contributing to the science that improves our understanding of life on Earth and to ecological literacy in civic life. Both are needed if we as a society are to make personal decisions and public policies to shape a healthier and more equitable future.

This book illustrates how the Bell Museum has remained steadfast in its service as the state's natural history museum. It also demonstrates how museums that adapt to the needs and challenges of the day continue to grow and thrive. No one can predict what the world will be like during the next 150 years, but there can be no doubt that the Bell Museum will play an important role in understanding the natural world and bridging connections between people and the planet.

Denise Young, Bell Museum Executive Director

Acknowledgments

This book is the result of the efforts of many individuals over a considerable period of time. We first started talking about documenting the museum's history in 2014, shortly after the funding for the new Bell Museum was finally secured. As preparations began for the move out of the Church Street museum, museum staff started the task of assessing more than seventy-five years of stored materials. As we sorted through reams of papers, publications, photographs, exhibit components, books, and other materials, we collated museum publications and readied materials to be sent to the archives at the University of Minnesota. We set aside and documented materials we knew would be useful in writing the story of the museum. Key in this endeavor was Dan Belden, whose organizational skills and dedication through the length of this project aided us all.

We thank the many people who agreed to be interviewed for the book. They are noted in "Selected References."

We Also Thank . . .

For support and advice—Denise Young, Bell Museum executive director, and the staff and advisory board members of the Bell Museum.

For funding support—grants funded by an appropriation to the Minnesota Historical Society from the Minnesota Arts and Cultural Heritage Fund. We thank the staff of the Minnesota Historical and Cultural Heritage Grants Program, especially program officer John Fulton. Additional grants were received from the Elmer L. and Eleanor J. Andersen Foundation and William C. Lawrence III.

For image archival research and acquisition help—Helen Gale Bruzzone; Katy Chayka, Minnesota Wildflowers; Brooke Doepke, marketing manager, Perkins and Will; Jane Greenberg; Carol Hall, Minnesota Department of Natural Resources; Edward "Ted" Hathaway, Hennepin County Library Special Collections; Carolyn and Larry Kuechle; Kristal Leebrick, *Park Bugle*; McGough Construction; Matthew Medler, collections manager, Cornell Lab of Ornithology; Doug Mock; Gary Nuechterlein and Deborah Buitron; Eva Nygards, curator of loans, Gothenburg Museum of Art; Gregory Raml, Special Collections Librarian, American Museum of Natural History; Larry Rudnick; Mr. and Mrs. Dell Schott; Christopher E. Smith, HerpMapper; Andria Waclawski, University of Minnesota; Richard Warner; Daniel Weddington, University of Michigan Special Collections Research Center; Kevin Winker, Curator of Birds, University of Alaska Museum of the North; and Adrienne Wiseman, Bell Museum. And, for technical support—Kyle Grindberg, Bell Museum.

For assistance with content—Ford Bell, Anita Cholewa, Susan Evarts, David Hartwell, Scott Lanyon, Sue Leaf, Jennifer Menken, Gordon Murdock, Lee Pfannmuller, Patrick Redig, Jennifer Stampe, and Rebecca Toov.

For graphic design concepts and support—Tim Eaton of Eaton & Associates Design Company and Kari Finkler of Eaton & Associates, now Kari Finkler Design.

We would especially like to thank our reviewers and the book's advisory group. The reviewers were Erik Moore, University archivist, University of Minnesota; John Moriarty, senior manager of wildlife, Three Rivers Park District; Jean O'Brien, Department of History, University of Minnesota; Mark Seeley, emeritus University of Minnesota Extension climatologist and meteorologist; and William Souder, author. Our advisory group was composed of Mark Borrello, Department of Ecology, Evolution, and Behavior, University of Minnesota; Susan Galatowitsch, Department of Fisheries, Wildlife, and Conservation Biology, University of Minnesota; Parke Kunkle, retired professor of astronomy, Minneapolis Community and Technical College; Sue Leaf, author; Lee Pfannmuller, retired Chief of Ecological Services, Minnesota Department of Natural Resources; Anna Pidgeon, Department of Forest and Wildlife Ecology, University of Wisconsin; Rob Silberman, Department of Art History, University of Minnesota; the late John Tester, Department of Ecology, Evolution, and Behavior, University of Minnesota; and Rebecca Toov, collections archivist, University of Minnesota.

Finally, we thank our partners and spouses—Gayle Thorsen, Ann Luce, Dan Engstrom, and Dale Kennedy—who lived with this book's creation for more time than they had imagined.

Appendix: The Bell Museum Dioramas

Caribou
Place depicted: Newfoundland (but similar to Red Lake peatland, Minnesota)
Date completed: 1911
Artist/preparator: Charles Corwin, Charles Brandler
Current status: elements in storage.

Dall Sheep
Place depicted: Kenai Peninsula, Alaska
Date completed: 1917
Artist/preparator: Charles Corwin, Charles Brandler
Current status: elements in storage.

White-tailed Deer
Place depicted: forest northeast of Duluth, Minnesota
Date completed: 1917
Artist/preparator: Charles Corwin, Charles Brandler
Current status: elements on display in Minnesota Journeys galleries and in storage.

Beaver
Place depicted: Itasca State Park, Minnesota
Date completed: 1919
Artist/preparator: Charles Corwin, Jenness and Olive Richardson
Current status: on display in Minnesota Journeys galleries.

Heron Lake
Place depicted: Heron Lake, Jackson County, Minnesota
Date completed: 1922
Artist/preparator: H. W. Rubens, Louis Agassiz Fuertes, Jenness and Olive Richardson
Current status: elements on display in Minnesota Journeys galleries and in storage.

Bears
Place depicted: near Beaver Bay, Minnesota
Date completed: 1925
Artist/preparator: Bruce Horsfall, Jenness and Olive Richardson
Current status: elements on display in Minnesota Journeys galleries and in storage.

Pipestone Prairie
Place depicted: Pipestone National Monument, Pipestone, Minnesota
Date completed: 1928
Artist/preparator: Bruce Horsfall, Jenness and Olive Richardson, Walter Breckenridge
Current status: elements on display in Minnesota Journeys galleries and in storage.

Wolves
Place depicted: Shovel Point, Tettegouche State Park on north shore of Lake Superior, Minnesota
Date completed: 1940
Artist/preparator: Francis Lee Jaques, Walter Breckenridge
Current status: on display in Minnesota Journeys galleries.

Snow Geese
Place depicted: Lake Traverse, Traverse County, Minnesota
Date completed: 1942
Artist/preparator: Francis Lee Jaques, Walter Breckenridge
Current status: on display in Minnesota Journeys galleries.

Lake Pepin
Place depicted: Sand Point, Frontenac State Park, Minnesota
Date completed: 1942
Artist/preparator: Francis Lee Jaques, Walter Breckenridge
Current status: on display in Minnesota Journeys galleries.

Sandhill Cranes
Place depicted: Agassiz Dunes Scientific and Natural Area, Minnesota
Date completed: 1946
Artist/preparator: Francis Lee Jaques, Walter Breckenridge
Current status: on display in Minnesota Journeys galleries.

Moose
Place depicted: Gunflint Lake, Boundary Waters Canoe Area Wilderness, Minnesota
Date completed: 1946
Artist/preparator: Francis Lee Jaques, Walter Breckenridge
Current status: on display in Minnesota Journeys galleries.

Elk
Place depicted: Inspiration Peak State Wayside, Otter Tail County, Minnesota
Date completed: 1948
Artist/preparator: Francis Lee Jaques, John Jarosz
Current status: on display in Minnesota Journeys galleries.

Tundra Swans
Place depicted: Long Meadow Lake, Minnesota River Valley National Wildlife Refuge
Date completed: 1948
Artist/preparator: Francis Lee Jaques, John Jarosz
Current status: on display in Minnesota Journeys galleries.

Big Woods (Maple–Basswood Forest)
Place depicted: Clear Lake, Maplewood Park, Waseca, Minnesota
Date completed: 1949
Artist/preparator: Francis Lee Jaques, John Jarosz
Current status: on display in Minnesota Journeys galleries.

Cascade River (Coniferous Forest)
Place depicted: upper falls of Cascade River, Cascade River State Park, Minnesota
Date completed: 1956
Artist/preparator: Francis Lee Jaques, John Jarosz
Current status: on display in Minnesota Journeys galleries.

Red Fox
Place depicted: southeastern Minnesota
Date completed: 1938
Artist/preparator: Walter Breckenridge
Current status: in storage.

Bald Eagle
Place depicted: near Canadian border, Minnesota
Date completed: 1941
Artist/preparator: Edward Brewer, Walter Breckenridge, John Jarosz
Current status: elements on display in Minnesota Journeys galleries.

Double-crested Cormorant
Place depicted: Lake of the Woods, Minnesota
Date completed: circa 1940s
Artist/preparator: Edward Brewer, Jenness Richardson, Walter Breckenridge
Current status: in storage.

Canada Lynx with Snowshoe Hare
Place depicted: northern Minnesota
Date completed: circa 1940s
Artist/preparator: Walter Breckenridge
Current status: on display in Minnesota Journeys galleries.

Grey Fox
Place depicted: Mississippi River bottomlands, Minnesota
Date completed: circa 1940s
Artist/preparator: Edward Brewer, Walter Breckenridge
Current status: on display at Church Street building.

Raccoon
Place depicted: no specific locale
Date completed: 1943
Artist/preparator: Edward Brewer, Walter Breckenridge
Current status: in storage.

Swallow-tailed Kite
Place depicted: Great River State Park, Minnesota
Date completed: 1946
Artist/preparator: Francis Lee Jaques, Walter Breckenridge
Current status: on display in Minnesota Journeys galleries.

American Badger with Pocket Gopher
Place depicted: sand dunes near Fridley, Minnesota
Date completed: 1949
Artist/preparator: Francis Lee Jaques, John Jarosz
Current status: in storage.

Marten and Fisher
Place depicted: northern Minnesota
Date completed: 1949
Artist/preparator: Francis Lee Jaques, John Jarosz
Current status: on display in Minnesota Journeys galleries.

Bobcat
Place depicted: Afton State Park, Minnesota
Date completed: 1951
Artist/preparator: Francis Lee Jaques, John Jarosz
Current status: in storage.

Otter
Place depicted: Minnehaha Falls, Minneapolis, Minnesota
Date completed: 1953
Artist/preparator: Francis Lee Jaques, John Jarosz
Current status: in storage.

Ring-necked Pheasant
Place depicted: North Oaks, Ramsey County, Minnesota
Date completed: 1953
Artist/preparator: Francis Lee Jaques, John Jarosz
Current status: in storage.

Loon
Place depicted: Basswood Lake, Boundary Waters Canoe Area Wilderness, Minnesota
Date completed: 1954
Artist/preparator: Francis Lee Jaques, John Jarosz
Current status: on display in Minnesota Journeys galleries.

Buffalo Fish and Western Grebe
Place depicted: Big Kandiyohi Lake, Kandiyohi County, Minnesota
Date completed: 1958
Artist/preparator: Francis Lee Jaques, John Jarosz
Current status: on display in Touch & See Lab.

Muskie, Northern Pike, Rock Bass
Place depicted: Mississippi River, above Minneapolis, Minnesota
Date completed: 1960
Artist/preparator: Francis Lee Jaques, John Jarosz
Current status: in storage.

Smallmouth and Largemouth Bass
Place depicted: no specific locale
Date completed: circa 1960s
Artist/preparator: Francis Lee Jaques, John Jarosz
Current status: in storage.

Passenger Pigeon
Place depicted: Minnesota River valley south of Minneapolis, Minnesota
Date completed: 1964
Artist/preparator: Francis Lee Jaques, John Jarosz
Current status: on display in Minnesota Journeys galleries.

Great Blue Heron
Place depicted: Rice Lake, Anoka County, Minnesota
Date completed: 1966
Artist/preparator: Edward Brewer, John Jarosz
Current status: in storage.

Coyote
Place depicted: St. Croix River valley, Minnesota
Date completed: 1967
Artist/preparator: Alfred Martin, John Jarosz
Current status: elements in storage.

Golden Eagle
Place depicted: Blue Mounds State Park, Minnesota
Date completed: 1968
Artist/preparator: Jerome Connelly, John Jarosz, Nan Kane
Current status: in storage.

Appendix

Selected Exhibitions at the Bell Museum

American Natural History Art Show, 1971

The Art of Francis Lee Jaques, 1972

Tony Angell—Portraits of Birds of Prey, 1975

Tropical Birds—Dana Gardner, 1977

Charles Harper Silkscreens, 1977

J. Fenwick Lansdowne Watercolors: Rails of the World, 1978

The Art of Scientific Illustration, 1978

Earth Imagery: Photographs by Richard Smith, 1979

Francis Lee Jaques, 1979

Artist-Naturalists of the Bell Museum, 1980

Birds of Prey: Louis Agassiz Fuertes, 1980

Pressed on Paper: Fish Rubbings and Nature Prints, 1981

Maynard Reece, 1982

Raptors in Art, 1982

Francis Lee Jaques: Artist-Naturalist, 1982

Where Our Birds Wintered, 1984

Viewing Nature with Electrons, 1984

For Peat's Sake: Researching Minnesota's Peatlands, 1984

A Celebration of Bats: The Photography of Merlin Tuttle, 1985

The Shell Game: Clam Fishing and the Pearl Button Industry, 1986

Understanding Life's Connections, 1986

The Chimpanzees of Gombe Stream, 1986

Walter J. Breckenridge: A Life in Natural History, 1987

The Birds of Minnesota, 1987

The Making of the Bell Museum Dioramas, 1987

Tricking Fish: How and Why Lures Work, 1987

AIDS and Intimate Choices, 1988

In the Realm of the Wild: The Art of Bruno Liljefors of Sweden, 1988

A Natural Treasury: The Building of the Bell Museum, 1988

The Language of Wood, 1989

Pioneers of Bird Illustration, 1989

Many Faces in Science, 1990

Contemporary Australian Aboriginal Art, 1990

Exploring Evolution, 1991

Embodied Spirits: The Art of the Asmat, 1991

Polar Images: The Art of David F. Parmelee, 1992

Exotic Aquatics: The Biology of Alien Infestations, 1992

The Art of the Wild and the Federal Duck Stamp Contest, 1993

Wildlife Art in America, 1994

Saving Endangered Species, 1995

Beauty and Biology: Butterflies and Moths in Art and Science, 1995

Peregrine Falcon: The Return of an Endangered Species, 1996

Owen J. Gromme, 1997

Margaret Mee: Return to the Amazon, 1998

Skulls! 1998

Morphin: The Science of Biological Change, 1998

Chased by the Light: Jim Brandenburg's 90-Day Journey, 1999

Wings of Paradise: The Great Moth Paintings of John Cody, 1999

Impressions of Nature: The Wildfowl Art of Frank W. Benson, 1999

Natural Wonders: Children's Art on the Environment, 2000

Life on the Edge—Alaska's Copper River Delta, 2000

Francis Lee Jaques: Master Artist of the Wild, 2001

Exploring Nature's Histories and Mysteries, 2002

Hidden World of Bears, 2002

The Lion's Mane, 2004

Birds in Art, 2005

Visions of Nature: The World of Walter Anderson, 2005

Bloom! Botanical Art through the Ages, 2006

Touch the Sky: Prairie Photographs by Jim Brandenburg, 2007

Project Art for Nature, 2007

Oddities and Curiosities of Nature, 2007

Behind the Diorama Glass, 2007

Mysteries in the Mud: Climate Change in the Big Woods, 2008

Paradise Lost: Climate Change in the North Woods, 2008

Art of the Wild and Minnesota Duck Stamp Artists, 2008

LIFE: A Journey through Time, 2009

Wolves and Wild Lands, 2009

Hungry Planet: What the World Eats, 2010

Drawn to Nature: Art Exhibition and Sale, 2009

The Shape of Nature: The Art of Francis Lee Jaques, 2010

Sustainable Shelter: Dwelling within the Forces of Nature, 2010

Preserving/Memory—Jeff Millikan, 2011

Natural Curiosities, 2012

Dig It! The Secrets of Soil, 2012

Freeze Frame: Capturing Nature in Winter, 2012

Audubon and the Art of Birds, 2013

No Man Is an Island—Sonja Peterson, 2013

Something Sketchy—Selections from Sketch Night, 2013

Birds and DNA, 2013

Eyes on the Universe, 2014

Impact: Birds in the Human-Built World, 2015

Stanley Meltzoff: The Great Sporting Fish, 2015

Peregrine Falcon: From Endangered Species to Urban Bird, 2016

A View from a Canoe: Paintings and Illustrations by Don Luce, 2016

Our Global Kitchen: Food, Nature, Culture, 2018

Weather to Climate, 2019

Wicked Plants, 2019

Audubon Animated, 2020

Appendix

Selected Publications of the Bell Museum

This list does not include external publications or articles for scientific journals or magazines written by museum personnel.

BELL MUSEUM OF NATURAL HISTORY LEAFLETS

No. 1. Blockstein, David. "New Thoughts on the Extinction of the Passenger Pigeon." 1984.

No. 2. Goodman, Billy, and John Slivon. "An Urban Eyrie for Minnesota Peregrines." No date.

No. 3. Tordoff, Harrison B. "A Peregrine Falcon Life Table." April 1986.

No. 4. Schmid, William D. "Lyme Disease." August 1986.

No. 5. Noetzel, David. "Tree Hole Mosquitoes and LaCrosse Encephalitis." January 1987.

No. 6. Krosch, Penelope. "Thomas Sadler Roberts and *The Birds of Minnesota*." April 1987.

No. 7. Herman, William S. "Monarch Migration and Life Cycle." May 1987.

No. 8. Nordquist, Gerda E., and Don Luce. "Bats in Minnesota." November 1987.

No. 9. Moriarty, John, and Don Luce. "Turtles in Minnesota." 1989. Revised 1997.

No. 10. English, Tom. "The Bell Museum's Double Elephant Folio." December 1989.

No. 11. Gerholdt, James E. Revised by Joan Galli and Carol Hall. Illustrations by Don Luce and Tom Klein. "Frogs and Toads of Minnesota." 1999.

BELL MUSEUM OF NATURAL HISTORY, OCCASIONAL PAPERS

No. 1. Roberts, Thomas S. *The Winter Bird-Life of Minnesota. Being an Annotated List of Birds That Have Been Found within the State of Minnesota during the Winter Months.* Minneapolis: Geological and Natural History Survey of Minnesota, 1916.

No. 2. Roberts, Thomas S. *The Winter Bird Life of Minnesota; The Migrations of Minnesota Birds; March and April Bird-Lore in Minnesota; May Bird-Lore in Minnesota: Four Radio Talks Broadcasted in the University of Minnesota Program over Station WCCO,* Minneapolis: University of Minnesota, 1926.

No. 3. Roberts, Thomas S. *Some Changes in the Distribution of Certain Minnesota Birds in the Last Fifty Years.* William Kilgore, *Breeding of the Connecticut Warbler* (Oporornis agilis) *with Special Reference to Minnesota.* Walter J. Breckenridge, *Breeding of Nelson's Sparrow* (Ammospiza nelsoni) *with Special Reference to Minnesota.* Walter J. Breckenridge, *A Hybrid Passerina* (Passerina cyanea + Passerina amoena). Minneapolis: University of Minnesota, 1930.

No. 4. Gunderson, Harvey L. *A Study of Some Small Mammal Populations at Cedar Creek Forest, Anoka County, Minnesota.* Minneapolis: University of Minnesota, 1950.

No. 5. Olson, Sigurd T., and William H. Marshall. *The Common Loon in Minnesota*. Minneapolis: University of Minnesota Press, 1952.

No. 6. Gunderson, Harvey L., and James R. Beer. *The Mammals of Minnesota*. Minneapolis: University of Minnesota Press, 1953.

No. 7. Underhill, James C. *The Distribution of Minnesota Minnows and Darters in Relation to Pleistocene Glaciation*. Minneapolis: University of Minnesota Press, 1957.

No. 8. Tester, John R., and William H. Marshall. *A Study of Certain Plant and Animal Interrelations on a Native Prairie in Northwestern Minnesota*. Minneapolis: University of Minnesota Press, 1961.

No. 9. Dickerman, Robert W. *The Song Sparrows of the Mexican Plateau*. Minneapolis: University of Minnesota Press, 1963.

No. 10. Phillips, Gary L., and James C. Underhill. *Distribution and Variation of the Catostomidae of Minnesota*. Minneapolis: Bell Museum of Natural History, 1971.

No. 11. Eddy, Samuel, Roy C. Tasker, and James C. Underhill. *Fishes of the Red River, Rainy River, and Lake of the Woods, Minnesota: with Comments on the Distribution of Species in the Nelson River Drainage*. Minneapolis: Bell Museum of Natural History, 1972.

No. 12. Moore, John W. *A Catalog of the Flora of Cedar Creek Natural History Area, Anoka and Isanti Counties, Minnesota*. Minneapolis: Bell Museum of Natural History, 1973.

No. 13. Birney, Elmer C., John B. Bowles, Robert M. Timm, and Stephen L. Williams. *Mammalian Distributional Records in Yucatan and Quintana Roo, with Comments on Reproduction, Structure, and Status of Peninsular Populations*. Minneapolis: Bell Museum of Natural History, 1974.

No. 14. Timm, Robert M. *Distribution, Natural History, and Parasites of Mammals of Cook County, Minnesota*. Minneapolis: Bell Museum of Natural History, 1975.

No. 15. Merrell, David J. *Life History of the Leopard Frog, Rana pipiens, in Minnesota*. Minneapolis: Bell Museum of Natural History, 1977.

No. 16. Weaver, Margaret G., and David J. McLaughlin, *Mushroom Flora of Minnesota: A Contribution*. Minneapolis: Bell Museum of Natural History, 1980.

No. 17. Hill, Suzanne Braun, and Dale H. Clayton. *Wildlife after Dark: Review of Nocturnal Observation Techniques*. Minneapolis: James Ford Bell Museum of Natural History, 1985.

No. 18. Karns, Daryl R. *Field Herpetology: Methods for the Study of Amphibians and Reptiles in Minnesota*. Minneapolis: James Ford Bell Museum of Natural History, 1986.

BELL MUSEUM OF NATURAL HISTORY, SELECTED TECHNICAL REPORTS

No. 1. Cochran, William W., Dwain W. Warner, and Dennis G. Raveling. *A Radio Transmitter for Tracking Geese and Other Birds*. Minneapolis: Museum of Natural History, University of Minnesota, 1963.

No. 2. Cochran, William W., and Edward M. Nelson. *The Model D-11 Direction Finding Receiver*. Minneapolis: Museum of Natural History, University of Minnesota, 1963.

No. 5. Birkebak, Richard C. *Measurement of the Thermal Radiation Properties of Biological Materials*. Minneapolis: Museum of Natural History, University of Minnesota, 1963.

No. 6. Tester, John R. *Radio Tracking of Ducks, Deer and Toads*. Minneapolis: Museum of Natural History, University of Minnesota, 1963.

No. 7. Cochran, William W., Dwain W. Warner, and John R. Tester. *The Cedar Creek Automatic Radio Tracking System*. Minneapolis: Museum of Natural History, University of Minnesota, 1964.

No. 9. Mech, L. David, and John R. Tester. *Biological, Behavioral, and Physical Factors Affecting Home Ranges of Snowshoe Hares* (Lepus americanus), *Raccoons* (Procyon Lotor), *and White-Tailed Deer* (Odocoileus virginianus) *under Natural Conditions*. Minneapolis: Museum of Natural History, University of Minnesota, 1965.

No. 11. Birkebak, Richard C., Eugene A. LeFebvre, and Dennis G. Raveling. *Estimated Heat Loss from Canada Geese for Varying Environmental Temperatures*. Minneapolis: Museum of Natural History, University of Minnesota, 1966.

OTHER PUBLICATIONS

Breckenridge, Walter J. *My Life in Natural History*. Compiled and edited by Barbara Breckenridge Franklin and John J. Moriarty. Minneapolis: Bell Museum of Natural History, University of Minnesota, 2009.

IMPRINT, publication of the Bell Museum for its members. December 1983/January 1984–Spring 2015.

Krosch, Penelope, ed. *Shotgun and Stethoscope: The Journals of Thomas Sadler Roberts*. Edited and transcribed by Penelope Krosch. Minneapolis: James Ford Bell Museum of Natural History, 1991.

Luce, Don. *Wildlife Art in America*. Minneapolis: James Ford Bell Museum of Natural History, University of Minnesota, 1994.

Luce, Don, Janet Pelley, and Kevin Williams. *The Making of the Bell Museum Dioramas*. Minneapolis: Bell Museum of Natural History, 1987.

Luce, Don, William Souder, Val Cunningham, and Gwen Schagrin. *Audubon and the Art of Birds*. Minneapolis: Bell Museum of Natural History, University of Minnesota, 2013.

Luce, Donald T., and Laura M. Andrews. *Francis Lee Jaques Artist-Naturalist*. Minneapolis: University of Minnesota Press, 1982.

Roberts, Thomas Sadler. *The Birds of Minnesota*. Minneapolis: University of Minnesota Press, 1932, 1936.

Roberts, Thomas S. *Annals of the Museum of Natural History, University of Minnesota, 1872–1939*. Minneapolis: University of Minnesota, 1939.

Self, Ruth Woolverton, and W. J. Breckenridge. *A Guidebook to the Minnesota Museum of Natural History at the University of Minnesota*. Minneapolis: University of Minnesota Press, 1952.

MOVIES AND TELEVISION

Bell *LIVE!* produced by the Bell Museum of Natural History: *Raptors LIVE!* 1994; *World of Wolves*, 1995; *Fire and the Forest*, 1996; *Aquatic Adventure*, 1998; *On the Prairie*, 1999; *Great Lakes: A Superior Adventure*, 2000; *Nature in the City*, 2001–2.

Breckenridge, Walter J. Selected films: *Wood Duck Ways*, c. 1940; *Minnesota Hawks and Owls*, c. 1945; *Exploring Eastern Minnesota's Waterways, Part I: The Southeast Hardwood Hill Country*, c. 1945; *Part II: The Lake Pepin Region*, c. 1957; *Spring Comes to the Subarctic*, 1955; *Island Treasures*, 1957; *Far Far North*, 1960s; *Birds of Minnesota Waterways*, 1960s; *Sand Country Wildlife*, date unknown.

Coffin, Barbara, executive producer; John Whitehead, director, writer, and editor; et al. *Minnesota: A History of the Land*, episodes 1–5. Minneapolis and St. Paul: Bell Museum of Natural History and Twin Cities Public Television, 2005–6. DVD.

Coffin, Barbara, Larkin McPhee, Steve Fisher, et al. *Troubled Waters: A Mississippi River Story*. Minneapolis: Bell Museum of Natural History, University of Minnesota, 2010. DVD.

Coffin, Barbara, and Don Luce, executive producers; Matt Ehling, producer. *Windows to Nature: Minnesota's Dioramas*. Minneapolis and St. Paul: Bell Museum of Natural History and Twin Cities Public Television, 2018. DVD.

Selected References

INTRODUCTION

Brataas, Anne, and James C. Underhill. "To Delight and Instruct: An Evolutionary Account of the First 100 Years of the Bell Museum of Natural History." Unpublished paper, 1980s.

Gruchow, Paul. *Grass Roots: The Universe of Home.* Minneapolis: Milkweed Editions, 1995, 212.

Minnesota Legislature. "An act to provide for a Geological and Natural History Survey of the State and entrust the same to the University of Minnesota." *General Laws*, chapter 30, 86–88, February 29, 1872 (approved March 1, 1872).

Olson, Manley. "MOU and the Bell Museum." Unpublished paper delivered at the Minnesota Ornithologists' Union Annual Meeting, December 2016.

Swanson, Gustav. "The Minnesota Ornithologists' Union." *The Flicker* 12, no. 4 (December 1940): 42–43.

Weller, Susan. "From the Director: Minnesota's State Natural History Museum: 140 years and Counting." *IMPRINT* (Spring 2012): 10.

DOCUMENTING MINNESOTA

Lass, William E. "Minnesota's Salt Lands Saga." *Minnesota History* 53, no. 1 (Spring 1992): 9–24.

Minnesota Legislature. "An act to provide for a Geological and Natural History Survey of the State and entrust the same to the University of Minnesota." *General Laws*, chapter 30, 86–88, February 29, 1872 (approved March 1, 1872).

Morey, G. B. "History of the Minnesota Geological Surveys and State Geologists." In *The State Geological Surveys: A History: A Project of the Association of American State Geologists*, by Arthur A. Socolow and the Association of American State Geologists. N.p.: The Association, 1988, 233–38.

Nachtrieb, Henry Francis. Papers. University of Minnesota Archives.

Roberts, Thomas S. *Annals of the Museum of Natural History, University of Minnesota, 1872–1939.* Minneapolis: University of Minnesota, 1939.

Underhill, James. "The Good Ship Megalops." *IMPRINT* 3, no. 2 (Spring 1986): 5–7.

Winchell, Newton Horace. *The Geological and Natural History Survey of Minnesota: The Ninth Annual Report for the Year 1880.* St. Peter: J. K. Moore, 1881.

THE MENAGE EXPEDITION

Broderick, Richard. "Lost Worlds: The Strange Fate of the Menage Expedition." *M: For Alumni and Friends*, Winter 1997, 8–9, 11.

Gale, Harlow. "Historical Sketch of the Minnesota Academy of Science." *Bulletins of the Minnesota Academy of Science* 4 (vol. listed as 1905; article dated 1906; may have been published 1910), 430–42.

"Of Scientific Interest: Results of the Menage Expedition to the Philippine Islands—Two Young Men Who, after Two Years of Work, Have Succeeded in Obtaining a

Valuable Collection for the Minnesota Academy of Natural Sciences—Rare Birds, Snakes, and Fish—The Largest Orang-Outang Ever Captured." *New York Times*, March 22, 1896, 26.

Timm, Robert M., and Elmer C. Birney. "Mammals Collected by the Menage Scientific Expedition to the Philippine Island and Borneo, 1890–1893." *Journal of Mammalogy* 61, no. 3 (August 1980): 566–71.

JOSEPHINE TILDEN

Bartell, Sheri L. *The History of the Department of Botany 1889–1989, University of Minnesota*. Minneapolis: University of Minnesota, 1999.

Brady, Tim. "Of Algae and Acrimony." *Minnesota Alumni Magazine* 107, no. 3 (January–February 2008): 30–33.

Horsfield, Margaret. "The Enduring Legacy of Josephine Tilden." *Hakai Magazine*, June 13, 2016. https://www.hakaimagazine.com/features/enduring-legacy-josephine-tilden/.

Moore, Erik A., and Rebecca Toov. "The Minnesota Seaside Station near Port Renfrew, British Columbia: A Photo Essay." *BC Studies: The British Columbian Quarterly* 187 (Autumn 2015): 291–302.

Pelley, Janet. "The Minnesota Seaside Station." *IMPRINT* 2 (Spring 1985): 5–7.

THOMAS SADLER ROBERTS

Breckenridge, W. J., and William Kilgore. "Thomas Sadler Roberts (1858–1946)." *Auk*, 63, no. 4 (October 1946): 574–83.

Krosch, Penelope. "Thomas Sadler Roberts and *The Birds of Minnesota*." James Ford Bell Museum of Natural History, leaflet no. 6, April 1987.

Krosch, Penelope, ed. *Shotgun and Stethoscope: The Journals of Thomas Sadler Roberts*. Edited and transcribed by Penelope Krosch. Minneapolis: James Ford Bell Museum of Natural History, 1991.

Leaf, Sue. Interview by Barbara Coffin. March 13, 2020, and April 26, 2020.

Leaf, Sue. *A Love Affair with Birds: The Life of Thomas Sadler Roberts*. Minneapolis: University of Minnesota Press, 2013.

Pfannmuller, Lee. Interview by Don Luce. February 2, 2020.

Roberts, Thomas Sadler. *The Birds of Minnesota*. 2nd ed. Minneapolis: University of Minnesota Press, 1936.

MAKING A MUSEUM FOR THE PUBLIC

Bell, James Ford. "Newfoundland Expedition Journal." Unpublished, 1909.

Leaf, Sue. *A Love Affair with Birds: The Life of Thomas Sadler Roberts*. Minneapolis: University of Minnesota Press, 2013.

Luce, Don, Janet Pelley, and Kevin Williams. *The Making of the Bell Museum Dioramas*. Minneapolis: Bell Museum of Natural History, 1987.

Roberts, Thomas S. *Annals of the Museum of Natural History, University of Minnesota, 1872–1939*. Minneapolis: University of Minnesota, 1939.

THE MANY TALENTS OF WALTER BRECKENRIDGE

Breckenridge, Walter J. *My Life in Natural History*. Compiled and edited by Barbara Breckenridge Franklin and John J. Moriarty. Minneapolis: Bell Museum of Natural History, University of Minnesota, 2009.

Crampton, Gayle, and Don Luce. "A Life in Natural History." *IMPRINT* 4, no. 1 (Winter 1987): 1–4.

Luce, Don. "Walter J. Breckenridge [In Memoriam]." *Minnesota Conservation Volunteer* 66, no. 390 (September–October 2003): 62–63.

Luce, Don. "A Naturalist's Eye, an Artist's Hand: Walter Breckenridge's Artworks Depict Minnesota's Natural Heritage." *Minnesota Conservation Volunteer* 72, no. 423 (March–April 2009): 40–49.

Roberts, Thomas S. *Annals of the Museum of Natural History, University of Minnesota, 1872–1939*. Minneapolis: University of Minnesota, 1939.

Tordoff, Harrison B. "In Memoriam: Walter John Breckenridge, 1903–2003." *The Auk* 121, no. 4 (October 2004): 1286–88.

EARLY PUBLIC EDUCATION

Breckenridge, W. J. "Naturalists for State Parks." *Minnesota Conservation Volunteer* 19, no. 111 (May–June 1956): 19–22.

Breckenridge, Walter J. *My Life in Natural History*. Compiled and edited by Barbara Breckenridge Franklin and John J. Moriarty. Minneapolis: Bell Museum of Natural History, University of Minnesota, 2009.

Cutler, Bruce Cutler (memories from wife, Lucy Cutler, interpretive guide in 1960s). Interview by Barbara Coffin. March 20, 2019.

Hansen, Barbara (Bell Museum interpretive guide in 1960s). Interview by Barbara Coffin. March 7, 2019.

Heig, Vince (Bell Museum graduate student and state park naturalist in 1950s). Interview by Barbara Coffin. March 19, 2019.

Leaf, Sue. *A Love Affair with Birds: The Life of Thomas Sadler Roberts*. Minneapolis: University of Minnesota Press, 2013.

Murdock, Gordon. "Bell History/Public Education: Thoughts from Gordon Murdock, Bell Museum Curator of Education 1982–2014." Unpublished paper, February 3, 2019.

"Richard Barthelemy." *IMPRINT* 5, no. 2 (Spring 1988): 4–6.

JAMES FORD BELL

Andrews, Roy Chapman, Frank Chapman, and T. S. Roberts. Correspondence. Archives of the American Museum of Natural History and University of Minnesota Archives.

Bell, Ford. Interview by Barbara Coffin and Don Luce. June 25, 2019, and February 11, 2020.

Bell, James Ford, and Family. Papers. Minnesota Historical Society, St. Paul.

James Ford Bell and His Books: The Nucleus of a Library. Minneapolis: Associates of the James Ford Bell Library, University of Minnesota, 1993.

Leaf, Sue. *A Love Affair with Birds: The Life of Thomas Sadler Roberts*. Minneapolis: University of Minnesota Press, 2013.

HEYDAY OF THE DIORAMAS

Allison-Bunnell, Steven. "Forward to Nature: Natural History Dioramas and the Idea of Wilderness." Paper presented at the Annual Meeting of the American Society for Environmental History, Tucson, Arizona, April 15–18, 1999.

Allman, Laurie. "Through the Looking Glass." *Minnesota Conservation Volunteer* 79, no. 469 (November–December 2016): 4–17.

Luce, Don. "Francis Lee Jaques: The Museum Diorama and Public Education." In *Value in American Wildlife Art, Proceedings of the 1992 Forum*. Edited by Peter Friederici and William V. Mealy. Jamestown, N.Y.: Roger Tory Peterson Institute of Natural History, 1992.

Luce, Don, Janet Pelley, and Kevin Williams. *The Making of the Bell Museum Dioramas*. Minneapolis: Bell Museum of Natural History, 1987.

Quinn, Stephen C. *Windows on Nature: The Great Habitat Dioramas of the American Museum of Natural History*. New York: Harry N. Abrams, 2006.

Schwarzer, Marjorie, and Mary Jo Sutton. "The Diorama Dilemma: A Literature Review and Analysis." Unpublished paper. The Oakland Museum of California, 2006.

Self, Ruth Wolverton, and W. J. Breckenridge. *A Guidebook to the Minnesota Museum of Natural History at the University of Minnesota*. Minneapolis: University of Minnesota, 1952.

Weller, Susan. "From the Director: Jaques' Dioramas: Our Minnesota Time Machine." *IMPRINT*, Spring 2014, 10.

Wonders, Karen. "The Illusionary Art of Background Painting in Habitat Dioramas." *Curator: The Museum Journal* 33, no. 2 (June 1990): 90–118.

TAKING FLIGHT

"F. L. Jaques, American Artist, 1887–1969." Special issue, *Naturalist* 21, no. 1 (Spring 1970).

Jaques, Florence Page. *Francis Lee Jaques: Artist of the Wilderness World*. Garden City, N.Y.: Doubleday, 1973.

Johnson, Dennis A. "The Shapes of Things." *Audubon Magazine* 85, no. 1 (January 1983): 66–83.

Johnson, Patricia Condon. *The Shape of Things: The Art of Francis Lee Jaques*. Afton, Minn.: Live Oak Press and Afton Historical Society, 1994.

Luce, Don. "The Legacy of Francis Lee Jaques." *IMPRINT* 18, no. 1 (Spring 2001): 6.

Luce, Donald T., and Laura M. Andrews. *Francis Lee Jaques Artist-Naturalist*. Minneapolis: University of Minnesota Press, 1982.

AT THE POLES

"*Antarctic Birds* by David F. Parmelee." *Signatures, Newsletter of the University of Minnesota Press*, Fall/Winter 1991.

Breckenridge, Walter J. *My Life in Natural History*. Compiled and edited by Barbara Breckenridge Franklin and John J. Moriarty. Minneapolis: Bell Museum of Natural History, University of Minnesota, 2009.

"David F. Parmelee: Scientist, Artist, Naturalist." *IMPRINT* 9, no. 2 (Spring 1992): 1–5.

Parmelee, David F. *Bird Island in Antarctic Waters: The Adventures of an Artist/Ornithologist on a Lonely Outcrop in the Far South Atlantic*. Minneapolis: University of Minnesota Press, 1980.

Parmelee, David F. *Antarctic Birds: Ecological and Behavioral Approaches*. Minneapolis: University of Minnesota Press, 1992.

Wilkie, R. J., and W. J. Breckenridge. "Naturalists on the Back River." Parts 1 and 2. *The Beaver: Magazine of the North*, Spring 1955, 42–45; Summer 1955, 9–13.

Winker, Kevin. "In Memoriam: David F. Parmelee, 1924–1998." *The Auk* 116, no. 3 (July 1999): 816–17.

THE BRIDE WORE . . . BOOTS?

Breckenridge, Walter J. *My Life in Natural History*. Compiled and edited by Barbara Breckenridge Franklin and John J. Moriarty. Minneapolis: Bell Museum of Natural History, University of Minnesota, 2009.

MIGRATIONS

"Of Birds, Rainforests, and Dwain Warner." *IMPRINT* 4, no. 1 (Winter 1987).

Warner, Richard. Interview by Lansing Shepard. August 2017.

Winker, Kevin, John H. Rappole, and Robert W. Dickerman. "In Memoriam: Dwain Willard Warner: 1917–2005." *The Loon* 77, no. 4 (Winter 2005): 191–94.

Winker, Kevin. "In Memoriam: Dwain Willard Warner: 1917–2005." *The Auk* 123, no. 3 (July 2006): 910–11.

TRACKING NATURE

Kuechle, Carolyn Ukura. *Animal Tracking Signals: Stories of the Development of Radiotelemetry in Minnesota*. St. Cloud: North Star Press of St. Cloud, 2009.

Tester, John (retired professor of ecology, evolution, and behavior, University of Minnesota). Interview by Barbara Coffin and Don Luce. June 14, 2014.

MYSTERY OF THE MISSING TOADS

Tester, John (retired professor of ecology, evolution, and behavior, University of Minnesota). Interview by Barbara Coffin and Don Luce. June 14, 2014.

Tester, John R. *Minnesota's Natural Heritage: An Ecological Perspective*. Minneapolis: University of Minnesota Press, 1995.

TOUCH AND SEE

Barthelemy, Richard E. "Our University's Museum of Natural History." *Minnesota Volunteer* 34, no. 199 (November–December 1971): 30–38.

Goodman, Billy. "An Oral History of the Touch and See Room." *IMPRINT* 5, no. 2 (Spring 1988): 3–6.

Johnson, Roger T., and Gordon Murdock. "The Touch and See Room." *IMPRINT* 11, no. 3 (Summer 1994): 4.

Menken, Jennifer (Touch & See Lab coordinator, Bell Museum). Interview by Lansing Shepard. July 9, 2019.

"Richard Barthelemy." *IMPRINT* 5, no. 2 (Spring 1988): 4–6.

PUBLIC PROGRAMS

Additional information on the history of the Bell Museum's public programs can be found in *IMPRINT*, the Bell Museum's publication for members, December 1983/January 1984 through Spring 2015; *IMPRINT's* calendar insert, 1984–2008; detailed public programs such as informal adult classes, field trips, family programs, summer camps, teacher workshops, lecture series, and special events.

Falk, John H., and Lynn D. Dierking. "The 95% Solution." *American Scientist* 98: 486–93.

Gilbertson, Donald. "From the Director: The Bell Museum's New Telecommunications Project." *IMPRINT* 7, no. 3 (Summer 1990), calendar insert.

"Island Expedition: Parkers Prairie Student Explores Hawaii with Team JASON." *IMPRINT* 18, no. 1 (Spring 2001): 10.

"JASON Team Trains 400 Teachers." *IMPRINT* 15, no. 1 (Spring 1998): 7.

Klaassen, Andrea. "Curating Community." *IMPRINT*, Spring 2012, 6.

Mattson, Shanai. "A Thirst for Science." *IMPRINT* 24, no. 3 (Fall 2007): 6–7.

Miller, Karin. "Peeking Out a Virtual Porthole: Jason Project." *MINNESOTA: The Magazine of the UMN Alumni Association* 95, no. 5 (May–June 1996): 16–17.

Murdock, Gordon (retired curator of education, Bell Museum). Interview by Barbara Coffin. February 25, 2019.

Nyberg, Katie (former Bell *LIVE!* coordinator, Bell Museum). Interview by Barbara Coffin. March 26, 2019.

Torgerson, Amy. "The Jason Project." *IMPRINT* 11, no. 3 (Summer 1994): 6.

INTERPRETING NATURE

Bell Museum of Natural History. "The Bell Museum Dioramas: Building an Ethic." *YouTube*. Uploaded by Bell Museum of Natural History, December 11, 2013. https://www.youtube.com/watch?v=HuGZXD554Kg.

Cummins, Heather (gallery programs coordinator, Bell Museum). Interview by Barbara Coffin. June 13, 2019.

Cutler, Bruce (memories from wife, Lucy Cutler, interpretive guide in 1960s). Interview by Barbara Coffin. March 20, 2019.

Hansen, Barbara (Bell Museum interpretive guide in 1960s). Interview by Barbara Coffin. March 7, 2019.

Heig, Vince (Bell Museum graduate student and state park naturalist in 1950s). Interview by Barbara Coffin. March 19, 2019.

Menken, Jennifer (Touch & See Lab coordinator, Bell Museum). Interview by Barbara Coffin. October 24, 2019.

Klaassen, Andrea. "Interpretive Guide Program Trains Educators." IMPRINT, Spring 2015, 6–7.

Teats, Brenda. "Guiding at the Bell Museum." IMPRINT 11, no. 3 (Summer 1994): 7.

FROM STUDENT GUIDE TO COLLEGE PROFESSOR

Pidgeon, Anna (professor, Department of Forest and Wildlife Ecology, University of Wisconsin). Interview by Don Luce, Gwen Schagrin, and Lansing Shepard. September 2018.

MAKING MOVIES

Amie, Jennifer. "Science Via Satellite: A Virtual Visit to a Vanishing Landscape." IMPRINT 16, no. 4 (Fall 1999): 2–4.

Bell Museum of Natural History. *Raptors LIVE!* Video. Minneapolis, 1994.

"Bell News: Teacher Workshops: Lessons in Sustainability." IMPRINT 24, no. 1 (Spring 2007): 9.

"Bell News: Ostroushko Concert Kicks Off Documentary Premiere." IMPRINT 23, no. 3 (Fall 2007): 8.

Breckenridge, Walter J. Papers. Bell Museum of Natural History Records (ua-00876). University of Minnesota Archives.

Coffin, Barbara, executive producer; John Whitehead, director, writer, and editor; et al. *Minnesota: A History of the Land*, episodes 1–5. Minneapolis and St. Paul: Bell Museum of Natural History and Twin Cities Public Television, 2005–6. DVD.

Coffin, Barbara, Larkin McPhee, Steve Fisher, et al. *Troubled Waters: A Mississippi River Story.* Minneapolis: Bell Museum of Natural History, University of Minnesota, 2010. DVD.

Coffin, Barbara, and Don Luce, executive producers; Matt Ehling, producer. *Windows to Nature: Minnesota's Dioramas.* Minneapolis and St. Paul: Bell Museum of Natural History and Twin Cities Public Television, 2018. DVD.

Lanyon, Scott. "From the Director: Virtual Reality: How Satellites and Cyberscience Fulfill a 19th-Century Mission." IMPRINT 16, no. 4 (Fall 1999): 9.

"Live via Satellite: 80,000 Kids Watch Science Unfold at a Minnesota Trout Stream." IMPRINT 15, no. 3 (Fall 1998): 10.

Nyberg, Katie (former Bell *LIVE!* coordinator, Bell Museum). Interview by Barbara Coffin. March 26, 2019.

"Public Television Documentary: *Minnesota: History of the Land* Launches to Great Acclaim." IMPRINT 22, no. 1 (Spring 2005): 9.

Sheperd, Nina, and Mark Cassutt. "U of M Bell Museum's New Film Programs—Documentary Studio, TV Special, Film Series Debut in October." *UM News Service,* September 21, 2006.

"TV Documentary Chronicles Minnesota's Landscapes and the People Who Change Them." IMPRINT 21, no. 4 (Winter 2005): 6.

HONEYBEES ON THE ROOF

Alfonso, Joe. "Final Report: Honey Bees and Humans: An Interdisciplinary 5th Grade Science Program." Unpublished report. Bell Museum of Natural History and Minneapolis Public Schools, 2014.

Amie, Jennifer. "The Plight of the Honeybee." *IMPRINT* 24, no. 2 (Summer 2007): 6–7.

"Bell Museum Bees." *IMPRINT* 24, no. 2 (Summer 2007): 6–7.

Kamesch, Hallie, and Lauren Sullivan. "Program Assessment: Honey Bees, Pollinators, and Food Program." Unpublished report prepared for the Saint Paul Public School System and the University of Minnesota Bell Museum, September 30, 2016.

Morrison, Deane. "Bees at the Bell: Children Explore the World of Bees at the Bell Museum." *UM News Service*, June 21, 2012.

Williams, Kevin (retired curator of education, Bell Museum). Interview by Barbara Coffin. July 10, 2019.

WIDENING THE INQUIRY

Askins, Robert (retired professor of biology, Connecticut College). Interview by Don Luce. June 21, 2019.

Barnwell, Frank (retired professor of ecology, evolution, and behavior, University of Minnesota). Interview by Don Luce and Gwen Schagrin. July 22 and July 29, 2019.

Barrowclough, George (curator of birds, American Museum of Natural History). Interview by Don Luce. July 29, 2019.

Bernstein, Neil (retired professor of biology, University of Iowa). Interview by Don Luce. July 2, 2019.

Blockstein, David (senior scientist of the National Council for Science and the Environment). Interview by Don Luce. July 2, 2019.

Corbin, Kendall (retired curator of systematics, Bell Museum, and professor of ecology, evolution, and behavior, University of Minnesota). Interview by Don Luce. July 10, 2019.

Cuthbert, Francie (professor of fisheries, wildlife, and conservation biology, University of Minnesota). Interview by Don Luce. July 8, 2019.

Genoways, Hugh H., Carleton J. Phillips, Jerry R. Choate, Robert S. Sikes, and Kristin M. Kramer. "Elmer Clea Birney: 1940–2000." *Journal of Mammalogy* 81, no. 4 (November 2000): 1166–76.

Lang, Jeff (emeritus professor of biology, University of North Dakota). Interview by Barbara Coffin and Don Luce. July 23, 2019.

Mock, Doug (professor of biology, University of Oklahoma). Interview by Don Luce. June 27, 2019.

Packard, Jane (emeritus professor of fisheries and wildlife, Texas A&M University). Interview by Don Luce. July 23, 2019.

Regal, Philip J. "Natural History in the Era of Genetic Engineering." *IMPRINT* 3, no. 4 (Fall 1986): 1–3, 8.

Smith, David (professor of fisheries, wildlife, and conservation biology, University of Minnesota). Interview by Don Luce. July 15, 2019.

Timm, Robert (emeritus professor of ecology, University of Kansas). Interview by Don Luce. June 26, 2019.

Wunderle, Joe (senior scientist, U.S. Fish and Wildlife Service). Interview by Don Luce. July 19, 2019.

NATURE VERSUS NURTURE

Buitron, Deborah, Gary Nuechterlein, et al. "McKinney's Grad Students' Remembrances." Unpublished compilation based on emails and telephone interviews by Don Luce. July 2017.

McKinney, Frank. "Autobiographical Notes." Unpublished document. 1986.

McKinney, Frank D. Information Files Collection (ua-01158). University of Minnesota Archives.

McKinney, Frank. "Comparative Approaches to Social Behavior in Closely Related Species of Birds." *Advances in the Study of Behavior* 8 (December 1978): 1–38.

McKinney, Frank "Animal Behavior: The Ethological Approach." *IMPRINT* 6, no. 4 (Fall 1989): 1–2.

McKinney, Frank, Lisa Guminski Sorenson, and Mark Hart. "Multiple Functions of Courtship Displays in Dabbling Ducks (Anatini)." *The Auk* 107, no. 1 (January 1990): 188–91.

McKinney, Frank, and Robert M. Zink. "Ornithology at the James Ford Bell Museum of Natural History and the University of Minnesota." In *Contributions to the History of North American Ornithology*, vol. 2 (Memoirs of the Nuttall Ornithological Club, no. 13), edited by William E. Davis Jr. and Jerome A. Jackson, 59–81. Cambridge, Mass.: Nuttall Ornithological Club, 2000.

Mock, Douglas W. "In Memoriam: Frank McKinney, 1928–2001." *The Auk* 119, no. 2 (April 2002): 507–9.

"William Brewster Memorial Award, 1994: Frank McKinney." *The Auk* 112, no. 1 (January 1995): 265–66.

MINNESOTA'S RAREST

Birney, Elmer C. "Why Study Bats in Minnesota?" *Minnesota Volunteer* 46, no. 270 (September–October 1983): 25–31.

Boman, Melissa. "*Eptesicus fuscus* (Big Brown Bat)." Rare Species Guide, Minnesota Department of Natural Resources website, 2018. https://www.dnr.state.mn.us/rsg/profile.html?action=elementDetail&selectedElement=AMACC04010.

Coffin, B. A., and Lee A. Pfannmuller. *Minnesota's Endangered Flora and Fauna*. Minneapolis: University of Minnesota Press, 1988, 473.

Eddy, Samuel, and James C. Underhill. *Northern Fishes with Special Reference to the Upper Mississippi River Valley*. 3rd ed. Minneapolis: University of Minnesota Press, 1974.

Gunderson, Dan. "Disease Continues to Batter Minnesota Bat Populations." MPR News website, posted March 28, 2019. https://www.mprnews.org/story/2019/03/28/disease-continues-to-batter-minn-bats.

Pfannmuller, Lee (coeditor *Minnesota's Endangered Flora and Fauna*). Interview by Barbara Coffin. May 27, 2020.

Phillips, Gary L., William D. Schmid, and James C. Underhill. *Fishes of the Minnesota Region*. Minneapolis: University of Minnesota Press, 1982.

Underhill, James C. "The Distribution of Small Stream Fishes in Minnesota: Intra-Specific Variability in the Common Shiner (*Notropis cornutus*), the Sand Shiner (*Notropis deliciosus*), and the Big-Mouth Shiner (*Notropis dorsalis*)." PhD diss., University of Minnesota, 1955.

FLIGHT OF THE PEREGRINE

Birney, Elmer C. "From the Director: Harrison B. Tordoff—Friend, Colleague, Mentor." *IMPRINT* 8, no. 3 (Summer 1991): 7–8.

Fallon, Jackie (state coordinator for Midwest Peregrine Society). Interview by Lansing Shepard. June 17, 2020.

Gill, Frank B. "In Memoriam: Harrison Bruce Tordoff, 1923–2008." *The Auk* 126, no. 2 (April 2009): 463–65.

Redig, Patrick (former director of Raptor Center, University of Minnesota). Interview by Lansing Shepard. June 2017.

Tordoff, Harrison Bruce, and Family. Papers. Minnesota Historical Society, St. Paul.

Tordoff, Harrison B. "Peregrine Falcons in Minnesota." *IMPRINT* 1, no. 1 (Winter 1984): 1.

Tordoff, Harrison B. "Almost Done!" *IMPRINT* 8, no. 3 (Summer 1991): 1–6.

Tordoff, Harrison B. "Trading Spaces: Peregrine Falcons' Home Improvement Strategies." *IMPRINT* 21, no. 3 (Fall 2004): 6.

Vezner, Tad. "Obituary, Harrison 'Bud' Tordoff." *St. Paul Pioneer Press*, July 24, 2008, B9.

ART AND NATURAL HISTORY

English, Tom. "The Bell Museum's Double Elephant Folio." James Ford Bell Museum of Natural History, leaflet no. 10, December 1989.

Luce, Don. "Natural History Art at the Bell Museum." IMPRINT 10, no. 1 (Winter 1993): 4–6.

Luce, Donald T. *Wildlife Art in America*. Minneapolis: James Ford Bell Museum of Natural History, 1994.

Luce, Don, William Souder, Val Cunningham, and Gwen Schagrin. *Audubon and the Art of Birds*. Minneapolis: Bell Museum of Natural History, University of Minnesota, 2013.

Peterson, Roger Tory. "Bird Art." *Natural History* 92, no. 9 (September 1983): 66.

Souder, William. *Under a Wild Sky: John James Audubon and the Making of The Birds of America*. New York: North Point Press, 2004.

Webster, William, and Byron Webster. "The American Museum of Wildlife Art." IMPRINT 10, no. 1 (Winter 1993): 1–3.

SCIENCE THROUGH THE LENS OF ART

Moen, Martin. "Experience Nature through Art." IMPRINT, Spring 2013, 5.

Moen, Martin. "Resident Artists Selected for 2015–16." IMPRINT, Spring 2015, 4.

CHANGE COMES TO THE "ETERNAL" MUSEUM

"Bell Museum News: AIDS and Intimate Choices Exhibit Wins Award." IMPRINT 6, no. 4 (Fall 1989), calendar insert.

"Bell Museum News: The Touring Exhibits Program Has Received a $50,000 Gift." IMPRINT 4, no. 3 (Summer 1987), calendar insert.

"Exhibits: *The Lion's Mane*." IMPRINT 21, no. 2 (Summer 2004): 5.

"Exploring Nature's Histories and Mysteries." IMPRINT 14, no. 1 (Spring 2002): 2–3, 8.

Gilbertson, Donald. "From the Director: A New Exhibit Hall Debuts with *In the Realm of the Wild, the Art of Bruno Liljefors of Sweden*." IMPRINT 5, no. 3 (Summer 1988): 8.

Griffiths, Alison. "Media Technology and Museum Display: A Century of Accommodation and Conflict." *Informal Learning Review* 66 (May–June 2004).

Korn, Randi. "An Analysis of Differences between Visitors in Natural History Museums and Science Centers." *Curator: The Museum Journal* 38, no. 3 (September 1995): 150–60.

McLean, Kathleen. *Planning for People in Museum Exhibitions*. Washington, D.C.: Association of Science-Technology Centers, 1993.

"Museum News: Museum Receives Grant to Begin Minorities in Science Exhibit." IMPRINT 6, no. 2 (Spring 1989), calendar insert.

Rader, Karen, and Victoria Cain. *Life on Display: Revolutionizing U.S. Museums of Science and Natural History in the Twentieth Century*. Chicago: University of Chicago Press, 2014.

St. John, Jetty. "Many Faces in Science." IMPRINT 7, no. 1 (Winter 1990), calendar insert.

COLLECTIONS OFFER CLUES TO ENVIRONMENTAL CHALLENGES

Baroni, Timothy J., and Andrew N. Miller. "MyCoPortal Helps Save Life!" Integrated Digitized Biocollections, iDigBio, posted. July 25, 2018. https://www.idigbio.org/content/mycoportal-helps-save-life.

Hemphill, Stephanie. "Deformed Minnesota Frogs Still Largely a Mystery 17 Years Later." MPR News website, posted July 17, 2012. https://www.mprnews.org/

story/2012/07/17/deformed-minnesota-frogs-still-largely-a-mystery-17-years-later.

Jansa, Sharon (curator of mammals, Bell Museum). Interview by Barb Coffin, Don Luce, Gwen Schagrin, and Lansing Shepard. September 25, 2019.

Kozak, Ken (curator of herpetology, Bell Museum). Interview by Barbara Coffin, Don Luce, Gwen Schagrin, and Lansing Shepard. October 22, 2019.

Mahoney, David. "Canaries in a Coal Mine: Solving the Puzzle of Minnesota's Deformed Frogs." *Annual Report 1999*. Minneapolis: Bell Museum of Natural History, 2000.

McLaughlin, David (former curator of fungi, Bell Museum). Interview by Barbara Coffin, Don Luce, Gwen Schagrin, and Lansing Shepard. September 25, 2019.

A BOTANICAL TREASURE

Adams, J. F. "Is Botany a Suitable Study for Young Men?" *Science* 9, no. 209 (February 4, 1887): 116–17.

Bartell, Sheri L. *The History of the Department of Botany 1889–1989, University of Minnesota*. Minneapolis: University of Minnesota, 1999.

Cholewa, Anita. "Brief History of the Herbarium." Unpublished document, c. 2015–16.

MacMillan, Conway. *The Metaspermae of the Minnesota Valley: A List of the Higher Seed-Producing Plants Indigenous to the Drainage-Basin of the Minnesota River*. Minneapolis: Harrison & Smith, 1892.

Moen, Martin. "Indispensible [sic] for 125 Years." *IMPRINT*, Spring 2014, 4.

University of Minnesota, Duluth. "Olga Lakela Herbarium." https://scse.d.umn.edu/olga-lakela-herbarium.

THE DNA REVOLUTION COMES TO THE BELL MUSEUM

Barrowclough, George, Joel Cracraft, John Klicka, and Robert Zink. "How Many Kinds of Birds Are There and Why Does It Matter?" *PLOS One* 11, no. 11 (November 23, 2016): 1–15. DOI: https://doi.org/10.1371/journal.pone.0166307.

Breckenridge Endowed Chair File. Endowed Chairs and Professorships. Information files collection (ua-01158). University of Minnesota Archives.

Breckenridge, Walter J. Papers. Bell Museum of Natural History records (ua-00876). University of Minnesota Archives.

Breining, Greg. "Robert Zink: Darwin's Finches and Sisyphean Evolution." *IMPRINT*, Spring 2014, 6–7.

Curtsinger, Jim (professor of ecology, evolution, and behavior, University of Minnesota). Interview by Don Luce. November 6, 2019.

Gilbertson, Donald. Letter to Douglas Pratt, Acting Dean of College of Biological Sciences, January 4, 1985, with enclosures from Kenneth Parks, George Watson, Francis James. Breckenridge Endowed Chair File, University of Minnesota Archives.

Gilbertson, Donald. "From the Director: The Breckenridge Chair in Avian Biology." *IMPRINT* 7, no. 2 (Spring 1990): 8.

Hill, Geoffrey E., and Robert M. Zink. "Hybrid Speciation in Birds, with Special Reference to Darwin's Finches." *Journal of Avian Biology* 49, no. 9 (September 2018).

"Katma Award 2015, to Bailey McKay and Robert Zink." *The Condor* 118, no. 1 (February 2016): 209–10.

Kessen, Ann (retired faculty, Century College). Interview by Don Luce. October 10, 2019.

Klicka, John (curator of birds, Burke Museum, University of Washington). Interview by Don Luce. October 23, 2019.

McKay, Bailey D., and Robert M. Zink. "Sisyphean Evolution in Darwin's Finches." *Biological Reviews* 90, no. 3 (August 2015): 689–98.

"A Report on the Breckenridge Chair in Ornithology." Unpublished report. University of Minnesota Foundation, 1993.

"William Brewster Memorial Award, 2005: Robert M. Zink." The Auk 123, no. 1 (January 1, 2006): 282–83.

Winker, Kevin (curator of birds, University of Alaska Museum of the North). Interview by Don Luce. October 23, 2019.

Zink, Robert (former Breckenridge Chair of Ornithology, Bell Museum). Interview by Don Luce. June 13, 2019.

Zink, Robert M., Hernan Vazquez-Miranda, and Frank Burbrink. "Species Limits and Phylogenomic Relationships of Darwin's Finches Remain Unresolved: Potential Consequences of a Volatile Ecological Setting." *Systematic Biology* 68, no. 2 (2019): 347–57.

RETHINKING THE TREE OF LIFE

Amie, Jennifer. "When Is a Tanager Not a Tanager? New Discoveries Will Shake Up Your Field Guide." *IMPRINT* 24, no. 3 (Fall 2007): 5–6.

Barker, Keith (curator of genetic resources, Bell Museum). Interview by Don Luce, Gwen Schagrin, and Lansing Shepard. September 3, 2019.

Barker, F. Keith, Kevin J. Burns, John Klicka, Scott M. Lanyon, and Irby J. Lovette. "New Insights into New World Biogeography: An Integrated View from the Phylogeny of Blackbirds, Cardinals, Sparrows, Tanagers, Warblers, and Allies." *The Auk* 132, no. 2 (April 2015): 333–48.

Barker, F. Keith, Alice Cibois, Peter Schikler, Julie Feinstein, and Joel Cracraft. "Phylogeny and Diversification of the Largest Avian Radiation." *Proceedings of the National Academy of Sciences of the United States of America* 101, no. 30 (July 27, 2004): 11040–45.

Berendzen, Peter B., Andrew Simons, and Robert M. Wood. "Phylogeography of the Northern Hogsucker, *Hypentieium nigricans:* Genetic Evidence for the Existence of the Teays River." *Journal of Biogeography* 30, no. 8 (August 2003): 1139–52.

Breining, Greg. "Seeing the Wren amongst the 'Little Brown Birds.'" *IMPRINT*, Winter 2012–13, 6–7.

Breining, Greg. "Re-writing Our Understanding." *IMPRINT*, Spring 2013, 2–3.

Breining, Greg. "Honey Badgers Don't Care: Bell Museum Scientists Study the Development of Genetic Resistance to Venom." *IMPRINT*, Spring 2015, 2–3.

Jansa, Sharon (curator of mammals, Bell Museum). Interview by Barbara Coffin, Don Luce, Gwen Schagrin, and Lansing Shepard. September 25, 2019.

Lanyon, Scott (former director of Bell Museum). Interview by Don Luce. January 30, 2020.

McLaughlin, David (former curator of fungi, Bell Museum). Interview by Barbara Coffin, Don Luce, Gwen Schagrin, and Lansing Shepard. September 25, 2019.

McLaughlin, David, David S. Hibbett, François Lutzoni, Joseph W. Spatafora, and Rytas Vilgalys. "The Search for the Fungal Tree of Life." *Trends in Microbiology* 17, no. 11 (2009): 488–97.

Morrison, Deane. "Of Possums and Pit Vipers: An Evolutionary Duel." *IMPRINT*, Spring 2012, 2–3.

"Reconstructing Trees: A Step by Step Method." Understanding Evolution, University of California Museum of Paleontology. https://evolution.berkeley.edu/evolibrary/article/0_0_0/phylogenetics_06.

Simons, Andrew (curator of fishes, Bell Museum). Interview by Don Luce, Gwen Schagrin, and Lansing Shepard. February 24, 2020.

BELL MUSEUM SCIENTISTS ON THE GLOBAL STAGE

Kozak, Ken (curator of herpetology, Bell Museum). Interview by Barbara Coffin, Don Luce, Gwen Schagrin, and Lansing Shepard. October 22, 2019.

Lanyon, Scott (former director of Bell Museum). Interview by Barbara Coffin, Don Luce, Gwen Schagrin, and Lansing Shepard. August 30, 2019.

National Science Foundation, Directorate for Biological Sciences, Division of Environmental Biology. "Assembling the Tree of Life (ATOL) Program Solicitation, NSF 10–513." National Science Foundation, revised November 2010. https://www.nsf.gov/pubs/2010/nsf10513/nsf10513.htm.

Reddy, Sushma (Breckenridge Chair of Ornithology, Bell Museum). Interview by Barbara Coffin, Don Luce, and Lansing Shepard. April 27, 2021.

Reddy, Sushma, Amy Driskell, Daniel L. Rabosky, Shannon J. Hackett, and Thomas S. Schulenberg. "Diversification and the Adaptive Radiation of the Vangas of Madagascar." *Proceedings of the Royal Society* B 279 (2012): 2062–71. DOI: https://doi.org/10.1098/rspb.2011.2380.

Simons, Andrew (curator of fishes, Bell Museum). Interview by Don Luce, Gwen Schagrin, and Lansing Shepard. February 24, 2020.

Stanton, Daniel (interim curator of lichens and bryophytes, Bell Museum). Interview by Barbara Coffin, Don Luce, and Lansing Shepard. April 28, 2021.

Yang, Ya (curator of plants, Bell Museum). Interview by Barbara Coffin, Don Luce, and Lansing Shepard. May 5, 2021.

BIODIVERSITY RESEARCH

"Bell News: Living Rainforest Exhibit Reveals University Research." *IMPRINT* 20, no. 2 (Summer 2003): 7.

Klaassen, Andrea. "Fig Findings from Down Under." *IMPRINT*, Spring 2015, 8.

Kozak, Ken (curator of herpetology, Bell Museum). Interview by Barbara Coffin, Don Luce, Gwen Schagrin, and Lansing Shepard. October 22, 2019.

Kozak, Kenneth H. "What Drives Variation in Plethodontid Salamander Species Richness over Space and Time." *Herpetologica* 73, no. 3 (September 2017): 220–28.

Lanyon, Scott. "From the Director: From Blackbirds to Fungi, Museum Science Documents Life's Variety." *IMPRINT* 21, no. 1 (Spring 2004): 10.

Maynard, Meleah. "Saving the Forest for the Trees." *Minnesota* (University of Minnesota Alumni Association) 107, no. 6 (July–August 2008): 14–19.

Morrison, Deane. "Museum Scientist's Discovery Lowers Estimates of Insect Species Found on Earth." *IMPRINT* 19, no. 3 (Fall 2002): 6.

Shepherd, Nina. "George's Jungle: An Effort to Save Some of the Planet's Most Critical Habitats Leads to Creating an International Research Station." *IMPRINT* 27, no. 1 (Summer 2010): 8–9.

Weiblen, George (science director, Bell Museum). Interview by Barbara Coffin, Don Luce, Gwen Schagrin, and Lansing Shepard. June 5, 2019.

Weiblen, George. "Plants and People in Papua New Guinea." *IMPRINT* 18, no. 2 (Summer 2001): 2–5.

ONE HUNDRED YEARS LATER

Bell Museum, University of Minnesota. Minnesota Biodiversity Atlas. https://bellatlas.umn.edu/.

Carlson, Bruce (supervisor of Minnesota Biological Survey, Minnesota Department of Natural Resources). Interview by Tim Whitfeld. September 5, 2019.

Converse, Carmen (retired supervisor of Minnesota Biological Survey, Minnesota Department of Natural Resources). Interview by Tim Whitfeld. September 12, 2019.

Minnesota Department of Natural Resources. *Field Guide to the Native Plant Communities of Minnesota. The Laurentian Mixed Forest Province.* Ecological Land Classification Program. Minnesota County Biological Survey and Natural Heritage and Nongame Research Program. St. Paul: Minnesota Department of Natural Resources, 2003.

Minnesota Department of Natural Resources. *Field Guide to the Native Plant Communities of Minnesota. The Eastern Broadleaf Forest Province.* Ecological Land Classification Program. Minnesota County Biological Survey and Natural Heritage and Nongame Research Program. St. Paul: Minnesota Department of Natural Resources, 2005.

Minnesota Department of Natural Resources. *Field Guide to the Native Plant Communities of Minnesota. The Prairie Parkland and Tallgrass Aspen Parklands Provinces.* Ecological Land Classification Program. Minnesota County Biological Survey and Natural Heritage and Nongame Research Program. St. Paul: Minnesota Department of Natural Resources, 2005.

Smith, Welby (botanist, Minnesota Biological Survey, Minnesota Department of Natural Resources). Interview by Tim Whitfeld. August 5, 2019.

Texler, Hannah (plant survey supervisor of Minnesota Biological Survey, Minnesota Department of Natural Resources). Interview by Tim Whitfeld. September 6, 2019.

Texler, Hannah. "The Big Reveal." *Minnesota Conservation Volunteer* 83, no. 489 (March–April 2020): 44–57.

COLLECTIONS GO ONLINE

Breining, Greg. "Making Collections Available to the World." *IMPRINT*, Spring 2014, 5.

Bell Museum, University of Minnesota. Minnesota Biodiversity Atlas. https://bellatlas.umn.edu/.

Zooniverse. Mapping Change. https://www.zooniverse.org/projects/zooniverse/mapping-change.

SAVING AN ENDANGERED MUSEUM

Birney, Elmer C. Papers. University of Minnesota Archives.

Birney, Elmer C. "Collegiate Priorities and Natural History Museums." *Curator* 37, no. 2 (1994): 99–107.

Bruininks, Robert (former president of University of Minnesota). Interview by Don Luce. November 13, 2019.

Genoways, Hugh H. "Challenges for Directors of University Natural Science Museums." *Curator* 42, no. 3 (July 1999): 216–29.

Hartwell, David (former chair of Bell Museum Advisory Board). Interview by Don Luce. December 19, 2020.

Hausman, Alice (Minnesota state representative). Interview by Don Luce. November 5, 2019.

Lanyon, Scott (former director of Bell Museum). Interview by Don Luce. January 3, 2020.

Lanyon, Scott M., Gordon Murdock, and Don Luce. "Planning for a Natural History Museum in a University Environment: A Case Study." *Curator* 43, no. 2 (April 2000): 88–92.

Moen, Marty (former director of communications, Bell Museum). Interview by Don Luce. October 30, 2019.

Pfannmuller, Lee (chair, Bell Museum Advisory Board). Interview by Don Luce. February 2, 2020.

Weller, Susan (former director of Bell Museum). Interview by Don Luce. November 11, 2020.

FROM THE EARTH TO THE COSMOS

Bonadurer, Robert (former director of Minneapolis Planetarium). Interview by Sally Brummel. October 2019.

Haarstick, Maxine. "How to Succeed in the Planetarium."

Museum News 43, no. 4 (December 1964): 17–21.

Kunkle, Parke (former board member, Minnesota Planetarium Society and Bell Museum). Interview by Sally Brummel. October 2019 and February 2020.

Mascotti, Lawrence (former program assistant, Minneapolis Planetarium, and former program committee member, Minnesota Planetarium Society). Interview by Sally Brummel. January 2020.

Morrison, Deane. "Dancing with the Stars." *Minnesota Alumni* 117, no. 4 (Summer 2018): 26–28.

Nelson, George "Pinky." "New Facility Brings Stars to Children." *Bemidji Pioneer*, February 28, 2002.

Nerdahl, Rodney (former program coordinator, Minneapolis Planetarium). Interview by Sally Brummel. October 2019.

Nerdahl, Rodney M. "With Stars in Our Eyes: A Fiftieth Anniversary Salute to the Planetarium." *Minneapolis Planetarium Update*, February, March, April 2001, 1–4.

Olson, Gail. "As Star Shows End, He Looks to the Future." *Northeaster* (Minneapolis), September 26, 2002, 1.

Rudnick, Lawrence (former board member, Minnesota Planetarium Society). Interview by Sally Brummel. January and February 2020.

Rudnick, Lawrence. "Minneapolis Planetarium, Status and Future Plans." Unpublished memo, 1998.

Schmickle, Sharon. "Minneapolis Planetarium Is No Longer Worlds Away." *Star Tribune*, April 3, 2005, 1B.

Shannon, Zella. "Minneapolis Planetarium." Unpublished report. Minneapolis Public Library and Information Center, 1984.

THE ROAD TO A REIMAGINED MUSEUM

"The Evolution of the Bell Museum." *IMPRINT* 14, no. 3 (Summer 1997): 7.

Illes, Molly. "Moose Exhibit Evaluation." Unpublished report. Bell Museum of Natural History, 2014.

Korenic, Mary S. "The Visitor and the Diorama at the Milwaukee Public Museum." Unpublished report. Milwaukee Public Museum, 1995.

Korsmo-Kennon, Peggy (former head of public programs, Bell Museum). Interview by Don Luce. January 2, 2020.

Krishtalka, Leonard, and Philip S. Humphrey. "Fiddling While the Planet Burns: The Challenge for U.S. Natural History Museums." *Museum News* 77, no. 2 (March–April 1998): 29–35.

Lanyon, Scott (former director of Bell Museum). Interview by Don Luce. January 3, 2020.

Lanyon, Scott. "Why Minnesota Needs a First-Rate Natural History Museum." Unpublished report. Bell Museum of Natural History, Minneapolis, 2000.

Luce, Don. "Biodiversity, Learning, and Exhibition in Natural History Museums: A Literature Review." Unpublished sabbatical report. Bell Museum of Natural History, Minneapolis, 2000.

Luce, Don, Barbara Coffin, Jim Roe, Shanai Matteson, and Bob Zink. "Minnesota's Nature Gallery: Design Document." Unpublished document. Bell Museum of Natural History, Minneapolis, 2009.

Marino, Margie, and Mary Fitzpatrick. "Bear in Mind—Visitors Like Dioramas! An Informal Study of the Denver Museum of Natural History Bear and Sea Mammal Hall." Unpublished report. Denver Museum of Natural History, 1998.

Miller, Holly. "Exploring Nature's Histories and Mysteries, Evaluation Report." Unpublished report, October 15, 2002.

"A Natural Evolution: Notes from a National Planning Workshop." Unpublished report. Bell Museum of Natural History, Minneapolis, June 2000.

People, Places & Design. "Public Expectation and

Perceptions about a Nature (Natural History) Museum Experience: A 'Front-end Study' for the New Bell Museum." Unpublished report: part 1, 2005; part 2, 2006; part 3, 2008.

Rudnick, Lawrence (professor emeritus, Minnesota Institute for Astrophysics). Interview by Don Luce. February 2020.

Schwarzer, Marjorie, and Mary Jo Sutton. "The Diorama Dilemma: A Literature Review and Analysis." Unpublished paper. The Oakland Museum of California, 2006.

Tirrell, Peter, and Robert Sullivan. "Report of the Survey Team for the American Association of Museums Museum Assessment Program III, Bell Museum of Natural History, University of Minnesota." Unpublished report, 1998.

"Visitor Experience/Interpretive Concept Plan, J. F. Bell Museum of Natural History and Planetarium." Unpublished document. October 2014.

Watson, Bill, and Shari Rosenstein Werb. "One Hundred Strong: A Colloquium on Transforming Natural History Museums in the Twenty-first Century." *Curator: The Museum Journal* 56, no. 2 (April 2013): 255–56.

Weller, Susan. "Director's Report." Unpublished report. Bell Museum of Natural History, Minneapolis, May 2015.

DESIGNING WITH NATURE

Dimond, Dave (architect and resilient design director, Perkins and Will). Interview by Barbara Coffin. January 29, 2020.

Lamb, Lynette. "Designed with Nature in Mind." *Minnesota Alumni* 117, no. 4 (Summer 2018): 24–25.

Pierce, Doug (architect and resilient design director, Perkins and Will). Interview by Barbara Coffin. January 21, 2020.

MOVING MINNESOTA

Brown, Terry (Museum Professionals). Interview by Don Luce. March 20, 2019.

Chase, Terry (Chase Studios). Interview by Don Luce. April 2020.

Coffin, Barbara, and Don Luce, executive producers; Matt Ehling, producer. *Windows to Nature: Minnesota's Dioramas*. Minneapolis and St. Paul: Bell Museum of Natural History and Twin Cities Public Television, 2018. DVD.

Emery, Megan (Midwest Art Conservation Center). Interview by Don Luce. March 25, 2020.

Jeffcoat, Kristy (Midwest Art Conservation Center). Interview by Don Luce. March 25, 2020.

Luce, Don. "Diorama Moving: Report on Planning Project." Unpublished report. Bell Museum of Natural History, Minneapolis, 1997; revised 2005.

Marquis, David. "Report on Condition and Adhesion of Bell Museum Diorama Backgrounds." Unpublished report, 2004.

THE EXPERIENCE

Bell Museum. "Interpretive Audio Tour." Unpublished audio file. Bell Museum, University of Minnesota, St. Paul, 2019.

Bell Museum Client Team and Gallagher and Associates. "Exhibition Script." Unpublished document, 2018.

Kroll, Mary, and the Bell Museum Client Team. "Overview—Exhibit Narrative of the New Bell Museum + Planetarium." Unpublished document, 2016.

Rosengren, John. "The Bell Comes Alive." *Minnesota Alumni* 117, no. 4 (Summer 2018): 16–23.

Contributors

Ford W. Bell, a veterinary oncologist, is the grandson of the Bell Museum's namesake, James Ford Bell. He was president of the American Alliance of Museums from 2007 to 2015 and has served on the boards of many museums, including the Bell Museum. He was a faculty member in the University of Minnesota College of Veterinary Medicine and was president of the Minneapolis Heart Institute Foundation.

Sally Brummel, the Bell Museum's planetarium manager, has led planetarium education and outreach for more than twenty years, producing fifty original programs nationally and internationally. She received a degree in physics from Albion College and a master's degree in education from the University of Minnesota.

Barbara Coffin was first associated with the Bell Museum in 1971 as a student tour guide. She went on to work for the Minnesota Department of Natural Resources and the Nature Conservancy before returning to the Bell Museum in 2004 as associate director of media productions and adult programs. She played an active role in the planning and design of the new museum's permanent exhibit galleries. She was the executive producer of the Emmy Award–winning television documentaries *Minnesota: A History of the Land*, *Troubled Waters: A Mississippi River Story*, and *Windows to Nature: Minnesota's Dioramas*. Throughout her career, she has been devoted to promoting ecological understanding of Minnesota's natural world through research, conservation, and educational outreach.

Don Luce, curator of exhibits at the Bell Museum, came to the museum in 1978. During his career of more than forty years, he curated most of the museum's temporary exhibitions, including *Exploring Evolution*, *The Lion's Mane*, *Wildlife Art in America*, and *Audubon and the Art of Birds*. He initiated the museum's traveling exhibitions program that circulated shows to venues throughout Minnesota and nationwide. An accomplished artist and scientific illustrator, he developed and expanded the museum's natural history art collection and related public programs. He is an authority on the history of dioramas and on the life and work of renowned diorama artist Francis Lee Jaques. He played a key role in the conception and design of Minnesota Journeys, the new museum's permanent exhibit gallery.

Gwen Schagrin has been associated with the Bell Museum since 1992, working in exhibits research, design, and production. Her experience and education as a librarian and graphic/interior designer enriched her contributions to several museum projects, including the Bell's *Wildlife Art in America* publication, and she assisted in the documentation and cataloging of the museum's wildlife art collection. She has expertise in the history, editions, and preservation of Audubon's *The Birds of America* and is a coauthor of the museum exhibition guidebook *Audubon and the Art of Birds*.

Lansing Shepard is a freelance writer who specializes in conservation, environmental policy, and natural history. His association with the Bell Museum began in the 1970s. Besides studying under several of the museum faculty who are profiled in this book, he wrote for the museum's *IMPRINT* publication, contributed to exhibition scripts, and coauthored the television documentary *Minnesota: A History of the Land*. A former reporter for the *Christian Science Monitor*, he is coauthor of *This Perennial Land: Third Crops, Blue Earth, and the Road to a Restorative Agriculture* and author of the *Northern Plains* volume of *The Smithsonian Guides to Natural America*.

George Weiblen, Bell Museum science director and curator of plants, is a Distinguished McKnight University Professor in the College of Biological Sciences, University of Minnesota. He studies the ecology and evolution of plants, pollinators, and pests in Papua New Guinea. He earned a doctorate in biology from Harvard University.

Tim Whitfeld, collections manager of the University of Minnesota Herbarium at the Bell Museum and the former director of the Brown University Herbarium, is an expert on the plants of Minnesota's forests, prairies, and wetlands. Studying tropical trees in Papua New Guinea, he earned a doctorate in plant biology from the University of Minnesota.

Denise Young, Bell Museum executive director since 2016, is a lifelong educator. She began her career as a kindergarten teacher in Durham, North Carolina, and moved from the classroom to the museum setting more than twenty years ago. Her doctorate in education is from the University of North Carolina–Chapel Hill.

Index

Compiled by Denise E. Carlson

All locations are in Minnesota unless otherwise indicated. Page numbers in italics indicate photographs, paintings, and other illustrations.

Abbe, Ernst, *157*

adaptations, 170, *171*, *222*

Akeley, Carl, taxidermy methods of, 35

albatross, *222*

Alfano, Joe, 112

algae, *20*, *171*; collections of, 151, 155; study of, 21, 23–24. *See also* Herbarium; lichens; plants

Amble, Tom, 210

American Badger with Pocket Gopher diorama, *235*

American Museum of Natural History (New York City), 35, 38, 48, 118, 135, 160; Akeley Hall of African Mammals, 58, *58*; Jaques' work at, 31, 52–53, *54*, *57*

American Museum of Wildlife Art, 138

amphibians: collections of, 151; curators of, 172, 175; study of, 2, 44, 86, 95, 115, 175–76. *See also* herpetology; reptiles; *and individual species*

Anderson, A. P., *15*

Anderson, James, 72

Anderson, Mike, 123

Anderson, Walter, wildlife art by, 138

Andrews, Robin, 102

Andrews, Roy Chapman, 52–53, *53*, *57*

animals: behavior of, 43, 45, 99, 114–18, 120–23; depicting, 28, 33–34, 55, 58–59, 62, 69; filming and recording, 28–29, 48, 140; fungi related to, *164*, *165*; interaction with the environment and plants, 34, *220*; radio-tagging, 82, *82*–*87*, *84*, *85*; rare and endangered, 29, 125–33; in Touch and See Room/Lab, 92, *92*. *See also* fauna; Minnesota Geological and Natural History Survey; wildlife; *and individual animal species*

art, integrating nature and science with, 3, 45, 69, 135–47. *See also* dioramas; exhibits; *and individual artists*

Askins, Bob, 116, 118

Atkinson, Frederick, 33

Audubon, John James: *The Birds of America*, 135, 137–35, *139*; *Carolina Parakeets*, 11, *134*; exhibitions of works by, 138, *139*

Back, George, 72

Back River expedition, 72, *72*, *73*

Baird, Thomas, 21, 23

Bakkom, Matthew, resident artist, 140

Bald Eagle diorama, *234*

Ballard, Robert, 95

Barker, Keith, *97*, *150*, 165–66, 169, 181, 182

Barnwell, Frank, 115

Barrowclough, George, 118

Barthelemy, Margaret (Richard's wife), 91

Barthelemy, Richard "Bart," 9, *51*, 99; hands-on learning programs, 2, 50–51, 90–93. *See also* Touch and See Room/Lab

Barzen, Jeb, 102

bats, on endangered species list, 126, *127*

Battistini, Anna Cerelia, resident artist, *143*

Bdote Learning Center, grand opening in 2018, 5

Bears diorama, 29, 39, *231*; moving, 210

Beauty and Biology (exhibit), *146*

Beaver diorama, 28, 34, *36–37, 35, 35, 38, 48, 222, 232*; moving, 214

behavior. *See* animals, behavior of; biology, behavioral; birds, behavior of

Bell, Ford, *186*

Bell, James Ford, 27, 33, *52,* 52–53, 55, 57, 58; donations from, 24, 29, 48, 52, 55; naming the museum, 57; purchases Heron Lake property, 35–38; work for new natural history museum building, 30–31, 55–56, 57. *See also* General Mills Corporation

Bell, James Stroud, 27, 28, 53, 55. *See also* Washburn-Crosby Milling Company

Bell *LIVE!,* 95, 106, *106,* 107. *See also* JASON Project

Bell Museum: addressing environmental issues, 94–98, 114–18, 137, 145, *147,* 150–53, 200, 203–3, 225; challenges faced by, 3, 186–89, 196, 197–98; changing names of the museum from 1872-2022, 6-11; connection to the University, 2–3, 145–46, 197, *196,* 198–100; founding, 1–2, 14; Minnesota Planetarium merger with, 4, 188–89, 190, 193–95, *192,* 199; mission, 2–4, 94, 96, 109, 196, 266; in Old Main, 1, 2, *2,* 15, *15,* 155; research at, 2–3, 5, 52–56; timeline, 6–11. *See also* General Museum; James Ford Bell Museum of Natural History; Minnesota Museum of Natural History; Museum of Natural History; Zoological Museum

Bell Museum, St. Paul Campus building: connecting visitors and scientists, 5, *196,* 198–100, *197, 227*; dioramas and exhibits in, 196–100, *196, 199,* 200, 217–27; Discovery Station, 200; Diversity Wall, 200; energy-efficient features, 203–3; exterior views, 5, *11, 189, 202, 204, 205, 206, 268*; grand opening, 5, 102, *187, 192*; groundbreaking, *186, 189, 192*; honeybee science lab, 112; Imagine the Future gallery, 200, *225*; interior views, *205, 217–27*; Life in the Universe gallery, *215–19*; Minnesota Journeys galleries, 200, *215–27*; Minnesota Planetarium in, *11,* 109, *189, 194,* 195; move to, *206,* 209–14, *210, 213*; planning and construction of, 186–89, *187*; Tree of Life gallery, *218*; Web of Life gallery, *220. See also* Touch and See Room/Lab; *and individual exhibits*

Bell Museum of Natural History/Minnesota Museum of Natural History, Church Street building, 5, *7, 100, 267*; art in, 135–43; dioramas in, 56–57, 99, *194*; Jaques Gallery, 137, 144; planning and construction of, *4,* 30–31, 52, 55–56, *57, 57*; West Gallery, 147. *See also* honeybee science lab (Honeybee Program); Touch and See Room/Lab

Bell Museum Productions, 107–9

Bell Museum Social, 97, 140, *142*

Bernstein, Neil, *117,* 118

Bewick, Thomas, 138; *Raven, 137*

Bieringa, Olive, resident artist, 140

Big Woods (Maple-Basswood Forest) diorama, 62, *62,* 63, 209, *207, 225, 235*; moving, *208, 214, 215*

Binning, Adele, 102

BioBlitz program, 97

biodiversity, 171, 200; research in, 150, 159, 169, 172–79, *173. See also* evolution; Minnesota Biodiversity Atlas; species

biology, 114–18, 152, 161; behavioral, 114; ecology of, 115; evolutionary, 118, 164–68, 169–71; exhibits on, *146,* 147; field, 74–75, 83, *105,* 115, 118; fish, 146; molecular, 3, 115, 164; wildlife, 83, 101. *See also* Lake Itasca Forestry and Biological Station (University of Minnesota); Minnesota Biological Survey (MBS); University of Minnesota, Animal Biology Building

birds: Antarctic and Arctic regions, 74, 75, *75*; behavior of, 44, 75, 79–80, *118*; bird-safe glass on St. Paul Campus building, 204; Central/South American, 82, *150, 166*; collections of, 80, *115,* 151, *161,* 170; curators of, 45, 75, 118, *170*; in dioramas, 38, 39, 61–62, 69; Mexican, 78–79, 80; paintings of, 46, *137,* 138; photographs of, *47,* 55, *141, 162, 163*; speciation of, 161–63, 166, 170–71, *171*; study of, 4–5, 25–31, 64. *See also* Breckenridge Chair of Ornithology; Minnesota Ornithologists' Union (MOU); Roberts, Thomas Sadler; *and individual paintings, photographs and species of birds*

Bird's Eye View of Climate Change, A (art installation), 142

Birke, Steve, *187*

Birney, Elmer, director Bell Museum, 10, 105, 115, *115,* 116, 186, *185*; work with endangered species, 125, 126

blackbirds, *166*

Blockstein, David, 116, 118

blue jay, 25

Bobcat diorama, 235

Boehnke, Luke, 210, 213

Boston Children's Museum, 51

botany, 15, 16, 156–59. *See also* Herbarium; plants

Boundary Waters Canoe Area Wilderness, 69

Bourns, Frank, and Menage expedition, 18, 19

Brandenburg, Jim, wildlife photography by, 138; *Touch the Sky Prairie,* 201

Brandler, Charles, 33

Brandon, Miranda, resident artist, 140, 141

Breckenridge, Dorothy Shogren (wife of Walter), marriage and honeymoon, 76, 76–77, 77

Breckenridge, Walter "Breck," 8, 42, 42–46, 44, 45; *American Kestrel,* 46; *The Birds of Minnesota* plates painted by, 136; Canadian toad project, 86, 86–87; dioramas prepared by, 44, 58, 59, 61–62, 64; director of Bell Museum, 2, 44–45, 62, 90; marriage and honeymoon, 76, 76–77, 77; nature films produced by, 43–44, 45, 45, 50, 72–73, 106–7, 107, 109; preparator at Bell Museum, 29, 39, 49; public programs launched by, 47, 50, 51; *Reptiles and Amphibians of Minnesota,* 44; on research expeditions, 72, 72–77, 74

Breckenridge Chair of Ornithology, 3, 5, 46, 160. *See also* birds; ornithology; Reddy, Sushma; Zink, Robert

Brewer, Edward, dioramas prepared by, 136

Brewer, Gwen, 122

Brewster, William, 26

Bright, Robert, work with endangered species, 125, 126

Brooks, Allan, *The Birds of Minnesota* plates painted by, 45, 136

Brown, Terry, 210, 208, 211, 213, 214, 212

Bruggers, David, 118

Bruininks, Robert, 108, 187, 188, 194

Brummel, Sally, 190, 193

Buettgen, Sue, 102

Buffalo Fish and Western Grebe diorama, 236

Buitron, Deborah, 123

Burger, Joanna, 117

Butler, Eloise, 156

Butler, Nathan, 26

butterflies, 138, 146, 147

Butters, Frederic, 157, 157

Cade, Tom, and return of peregrine falcons, 129–30

Café Scientifique, 96, 97

Canada Lynx with Snowshoe Hare diorama, 234

Caribou diorama, 6, 32–33, 33, 34, 55, 230

Carlos Avery Wildlife Management Area, 118

Carson, Rachel, *Silent Spring,* 114

Cascade River (Coniferous Forest) diorama, 8, 62, 63, 64, 144, 234; moving, 214

catbird, 47

Catesby, Mark, wildlife paintings, 138

Catlin, George, Pipestone prairie painting, 38–39

Cedar Creek Natural History Area (University of Minnesota), 45, 74, 75, 118, 122, 200; radio telemetry project, 82–85, 84, 115, 121

Chapman, Frank, 38, 53, 57, 66

Chase, Terry, 92, 210, 211, 212

children. *See* education; honeybee science lab (Honeybee Program); Touch and See Room/Lab

Cholewa, Anita, 158, 158

Church, Frederic, 34

citizen scientists, 156, 173, 174, 175, 182, 225. *See also* science

climate change, 125, 126, 140, 171, 176, 204; exhibit discussing, 142. *See also* conservation; environment, the

Cobb, John, 26

Cochran, William, 82, 83, 84

Coffin, Barbara, 107, 108, 198

Collecting Memory: Photographs and Installations (exhibit), 142

collections: educational uses, 47, 90–91, 135–43; growth of, 16, 30, 77; natural history, 14, 15, 19, 21, 52, 114, *218*; online, 181–82; preparation of, *115*, *180*; research uses, 150–53; scientific, 95. *See also* algae, collections of; birds, collections of; dioramas; exhibits; fish, collections of; fungus/fungi, collections of; Herbarium; mammals, collections of; Minnesota Geological and Natural History Survey; mollusks, collections of; plants, collections of; reptiles, collections of

Coniferous Forest diorama. *See* Cascade River (Coniferous Forest) diorama

conservation, 38, 39, 53, 176, 182; St. Paul Campus building's energy-efficient features, 38, 203–3; Thomas Roberts's involvement in, 30, 135. *See also* climate change; environment; habitat loss; species, conservation of

Corbin, Kendall, director of Bell Museum, 10, 115, *116*, 118

Corwin, Charles, dioramas prepared by, 35, *38*, 136

Coues, Elliott, 26

Coyote diorama, *237*

cranes. *See* sandhill cranes; Sandhill Cranes diorama

Cronquist, Arthur, 157

Curtsinger, Jim, 160

Cushing, Edward, 179

Cuthbert, Francie, 116

Dakota peoples: language, 222; location of Bell Museum on land of, 5, 266

Dall Sheep diorama, 34, 55, 210, *230*

Dart, Leslie, 27, *27*

Darwin, Charles, natural selection theory, 162, 164. *See also* evolution

Dayton, Mark, Governor, approves funding for new Bell Museum, 188, 189, 195

Dayton, Wallace, donations from, 160

DDT, 129, *130*. *See also* pesticides

Deane, Ruthven, 135

Delta Waterfowl Research Station (Manitoba, Canada), *120*, 121

Densmore, Mabel, 27, *28*, 30

Diamond, Jared, speech at Bell Museum, 94

Dickie, Trevor, 210

Dill, Homer, 42

Dimond, David, 203

dioramas, 3, 33–39, 58–65, *230–38*; in Church St. building, 56–57, 99, *194*; circulating to schools, 42–43, 49, *49*; educational uses, 42–43, 49, *49*, 95, 99, 135–36; moving, 209–14, *208*, *211*, *212*, *213*; mural paintings in, 2, 52–53, *54*, 210, *209*, 213, 214; with Native languages, *220*; preparation of, *34*, 35–39, *39*, 42, 49, 53, 55, 59, 135–36, *207*; transformation of, 140, *141*, 147, 196–100, *195*; videos enhancing, 108, 109; visitors interacting with, 51, *51*, 224. *See also* exhibits; Jaques, Francis Lee, diorama mural paintings; *and individual dioramas*

diversity. *See* biodiversity

DNA sequencing, 160–68, *161*, *168*, 169–70

Dobzhansky, Theodosius, 164

Double-crested Cormorant diorama, *44*, 236

ducks, 46, *83*, 121, *121*, 122; radio-tagging, 85

Dudley, Ian, 205

ecology, 34, 114–18, 135, 175; avian, 82, 105; migration, 79–80; studying, 74, 83, 101, 102; Waubun Prairie, 86–87

education: collections' role in, 47, 90–91, 135–43; curators of, 94, 95, 105, 110; dioramas' uses for, 42–43, 49, *49*, 95, 99, 135–36; goal of inclusivity, 3–4; graduate museum studies program, 95, 116–18, 123; hands-on learning, 2, *90*, 90–93, *91*, *92*; in natural history museums, 48–50, 92, 94; nature films, 43–44, *45*; public programs, 47–51, *94*, 94–98, *95*, 99–105, 186, 197, 199; role playing, 99, 101, *101*; school groups, 48, 50–51, *51*, 65, 95, 98, *101*, *102*; Schott Learning Kit program, 95; science, 110–12; St. Paul Campus building's learning landscape, 204–3; summer camps, 95, *95*, *103*, 105; Sunday afternoon programs, 28, 44, *45*, 49, *49*, 50; videos enhancing, 108–9. *See also* Bell LIVE!; exhibits; guides, interpretive; honeybee science lab (Honeybee Program); JASON program; Touch and See Room/Lab

Egge, Jacob, *168*, *198*

egrets, *122*

Elk diorama, 62, 99, *221*, 235

Emery, Megan, 210, 213

Endangered Species Acts: Minnesota (1981), 125; U.S. (1972), 84, 129, 162. *See also* animals, rare and endangered; plants, rare and endangered

entomologists/entomology, 16, 111, 147, 188

environment, 29, 30, 34, *147, 222*. *See* Bell Museum, addressing environmental issues; Bell Museum, St. Paul Campus building, energy-efficient features; climate change; conversation; habitat loss

equity and inclusion, viii. 3–4, 196

Erickson, Al, on Antarctic expedition, 74

ethology, 45, 120, 121. *See also* animals, behavior of; McKinney, Frank

Evans, Kory, *151*

Evarts, Sue, 123

evolution, 66, 114–18, 121, 165; biology of, 118, 164–68, 169–71; Darwin's theory of, 162, 164. *See also* biodiversity; DNA sequencing; Tree of Life Project

Evolution Study Group, 115

exhibits: art, 46, 137–35; curators of, 92, 198, 209; in St. Paul Campus building, 196–100, *196, 199, 200*; temporary and traveling, 144–47, *145, 146, 147*; videos enhancing, 108–9; visitors interacting with, 51, *51, 146, 224, 227*. *See also* dioramas; Touch and See Room/Lab; *and individual exhibits*

Exotic Aquatics (exhibit), *146*

Exploring Nature's Histories and Mysteries (exhibit), *196, 199, 197, 200*

Falls Creek Scientific and Natural Area (SNA), 178, 179

fauna: endangered, 125–33; Ice Age, 203; Minnesota and Midwest, 72, 178. *See also* animals; wildlife

Field Museum (Chicago), 19, 33, 35, 49, 187, 196

films, nature, 106–9. *See also* Breckenridge, Walter, nature films produced by; Roberts, Thomas Sadler, nature films produced by

finches, Darwin's, 162, *163*

fish: collections of, 19, 151, *151*, 152; curators of, 125, 167–68; distribution of, 167–68; exhibit on, 146, *146*; study of, 16, *17*, 168

flora: endangered, 125–33; Minnesota and Midwest, 72, 155, 157–58, 178. *See also* Herbarium; plants

Folwell, William W., 2; vision for Bell Museum, 2–3; writes Geological and Natural History Survey Act of 1872 (Minnesota), 1, 14

Forester, James, 109

Fraser, Don, support for planetarium, 192

frogs, deformed, 152–53, *153*

Fuertes, Louis Agassiz: *The Birds of Minnesota* plates paintings, 136; Heron Lake diorama paintings, 38

fungus/fungi, 126, *166*; animals related to, *164*; collections of, 151, 155, *156, 165*. *See also* algae; Herbarium; mushrooms

Garrigan, Mary Beth, 102

Gaylord, Edison, 26

General Mills Corporation, 52. *See also* Bell, James Ford; Washburn-Crosby Milling Company

General Museum, 6, 15, *15*, 155. *See also* Bell Museum; James Ford Bell Museum of Natural History; Minnesota Museum of Natural History; Museum of Natural History; Zoological Museum

gene sequencing. *See* DNA sequencing

genetic resources, curators of, 150, 165, 181

Geological and Natural History Survey Act of 1872 (Minnesota), 1, 14, 126. *See also* Minnesota Geological and Natural History Survey; natural history

Gilbertson, Don, 95, 147, 187; director of Bell Museum, 9, 160, 186

gnatcatchers, California, 162, *163*

Golden Eagle diorama, 144, *237*

Goodall, Jane, speech at Bell Museum, 94, *94*

Goodwin, Chris, 102

goose, radio-tagging, *82*

Great Blue Heron diorama, *236*

grebes, *123, 238*

Green, Robert and Beryl, Arctic expedition, 76, *77*

Grey Fox diorama, *234*

Grinnell, George Bird, 26

Gruchow, Paul, 1, 125

guides, interpretive, 51, 65, 92, 98; student program, 99–105, *101, 102, 103*

Gunderson, Harvey, 45, 50, 72

Haarstick, Maxine, 8, 191, *189*, 195

habitat groups. *See* dioramas

habitat loss, 30, 79, 129, 172, 179, 195, 196. *See also* conservation; environment, the; species, loss of

Hadland, Curt, 147

Hall, Carol, *Amphibians and Reptiles in Minnesota*, 44

Halvorson, Joel, 193, 195

Hansen, Barbara, 102

Hartwell, David, *186*

Hasselmo, Nils, 187

Hatch, Jay, work with endangered species, *126*

Hausman, Alice, *186*, 188–89

hawks, 44

Herbarium, 24, 151, 153, 155–59, 165; collections in, *10, 154, 158, 159, 179, 182*. *See also* algae; botany; flora; fungus/fungi; plants

Heron Lake diorama, 29, 35–38, *39, 233*

herpetology, 44, 115, 163, 175. *See also* amphibians, study of; reptiles, study of

Herrick, Clarence, 15, 26

honeybee science lab (Honeybee Program), 4, *10*, 97, *110*, 110–13, *111, 112, 113*

Horsfall, Bruce, dioramas prepared by, 39, 42, 136

Hungry Planet (exhibit), 147

IMPRINT newsletter, *10*, 95, 186

Indigenous peoples, viii, 1, 5; collaborating with Bell scientists, 4, *173, 174*, 175; as expedition guides, *18*, 33; language in diorama interpretation, 222. *See also* Dakota peoples; Native Americans

Inuit peoples, 73, *73*

Isua, Brus, *174*

Itasca State Park, 35, 50, *50*

James Ford Bell Museum of Natural History, 9. *See also* Bell Museum; General Museum; Minnesota Museum of Natural History; Museum of Natural History; Zoological Museum

Jansa, Sharon, 152, 166–67, *167*, 169

Jaques, Florence Page (wife of Francis), 66, 68, 137; *Canoe Country*, 69, *69*

Jaques, Francis Lee, 66, *66*–70, *68*; *The Birds of Minnesota* plates paintings, *69*, 136–37; diorama mural paintings, 2, 31, 58–59, *59*, 61–62, *63, 64*, 203, 209, *209*; exhibition of works, *144*, 144–45; wildlife paintings, *67, 70, 136*, 137; work at American Museum of Natural History, 31, 52–53, *54, 57*

Jarosz, John: on Arctic research expedition, 72, *72, 73*; dioramas prepared by, 62, *62, 64, 207*

JASON Project, 95, *96*, 106. *See also* Bell LIVE!

Jeffcoat, Kristy, 210, 213, *211*

Johnson, Roger, 91

Karns, Daryl, 116; work with endangered species, 125

Keogh, Sean, work with endangered species, *126*

Kilgore, William, 27, 29, *48*

Kimball, Jim, on Arctic research expedition, 73

Kirkwood, Sam, 115

Klicka, John, 161–62

Komperud, Sarah, *191*

Kottler, Malcolm, 115

Kozak, Ken, 169, 172, 175–76, *177*

Kuechle, Larry, 83, 84, 85, *85*

Kunkle, Parke, 192

Lake Itasca Forestry and Biological Station (University of Minnesota), 23, 74, 75, 115, 118

Lakela, Olga, *30*, 156, 156–57

Lake Pepin, *17*

Lake Pepin diorama, 31, 58, 232

Lang, Jeff, work with endangered species, 125

Lank, David, 137

Lanyon, Scott, 109, 165–66, *166*; director of Bell Museum, *108*, 169, 181, 187, *185*, 188, 196, 197

Larson, Shannon, *212*

Lee [Professor], *15*

Lewis, Don, *50*

lichens, 151, 156, 171, 179, 181. *See also* algae; mosses

Liljefors, Bruno: *Foxes*, 137; wildlife art of, 137, 147

Lion's Mane (exhibit), 199, *197*

Llano, George, *74*

Loon diorama, *236*

Lorenz, Konrad, *120*

Luce, Don, *139*, *144*, 144–45, 198, 209

Lutter-Gardella, Christopher, resident artist, 142

MacMillan, Conway, *15*, 16, 21, 23, *155*, 155–56

mammals: African, 58, *58*; collections of, 2, 19, 34, 42, 52, *115*, 151, 155, *213*; curators of, 45, 74, 105, 125, 152, 166, *167*, 186; studying, 44, 87, 123

mammoth, in St. Paul Campus building, *200*, *203*, *219*

Many Faces of Science (exhibit), *4*, 146

Mapping Change project, *180*, 182

Marten and Fisher diorama, *235*

Matteson, Shanai, *96*, *97*

Mayr, Ernst, 66, 115, 166

McKay, Bailey, 162–63

McKinney, Frank, 45, 116, *120*, 120–23, *121*, *122*, 160

McLaughlin, David, 165, *165*

Mech, David, 118

Mee, Margaret, botanical art by, 138

Menage, Louis F., 19

Menage expedition, *18*, 19

Menken, Jennifer, 93, *93*, 101

Midwest Peregrine Restoration Project, 130

Mierow, Dorothy, *62*

Miller, Tom, *27*

Millikan, Jeff, resident artist, 142; *He Had Spent His Lifetime Gathering Eggs*, *140*

Mima mounds, 86, *87*

Mindoro dwarf buffalo, *18*

Minneapolis Planetarium. *See* Minnesota Planetarium, Minneapolis Public Library location

Minnesota: A History of the Land (television series), 107–8, *108*, 109

Minnesota Academy of Natural Sciences, 19

Minnesota Biodiversity Atlas, 178–79, 181–82, *182*

Minnesota Biological Survey (MBS), 16, 158, *178*, 178–79

Minnesota Geological and Natural History Survey, 1, 14–17, *15*, *17*, 26, 28, 126, 155; updating, 178–79. *See also* botany; Geological and Natural History Survey Act of 1872 (Minnesota); natural history; zoology

Minnesota Museum of Natural History, 7, 57, 78, 86, 99. *See also* Bell Museum; General Museum; James Ford Bell Museum of Natural History; Museum of Natural History; Zoological Museum

Minnesota Ornithologists' Union (MOU), *4*, 4–5, 160. *See also* ornithology

Minnesota Planetarium, 190–95; ExploraDome traveling program, *190*, *193*, 193–94, *195*; merger with Bell Museum, 4, 188–89, 190, 193–95, *192*, 199; Minneapolis Public Library location, *188*, 190–93; shows for, 109, *109*, 194; Spitz projectors, 190, *191*, *189*

Minnesota Seaside Station (Vancouver Island, Canada), 22, *23*, *23*, 156

Mock, Doug, 116, 117, 123

mollusks: collections of, 19, 151, 155; curators of, 125, 126. *See also* mussels

Momeni, Ali, resident artist, work on dioramas, *141*

Moose diorama, *51*, *64–65*, *101*, *222*, *234*; moving, *206*, 213; reinterpreting, 198, *196*

Moran, Thomas, 34

Moriarty, John, *Amphibians and Reptiles in Minnesota*, 44

Moss, Gary, *Family of Swans*, *10*, 138

mosses, 33, 151, 156, 171, 214. See also lichens

moths, 138, *146*, 147

Murdock, Gordon, 94–95, *95*, 105

Murray, Kate, 104

Museum of Natural History, 7. See also See also Bell Museum; General Museum; James Ford Bell Museum of Natural History; Minnesota Museum of Natural History; Zoological Museum

museums, definition of, 1. See also natural history museums; *and individual museums*

mushrooms, 151, 153, *153*. See also fungus/fungi

Muskie, Northern Pike, Rock Bass diorama, *236*

mussels, 126, *126*, 152. See also mollusks

Mysteries in the Mud (exhibit), 108, *109*, 199

Nachtrieb, Henry, director of Bell Museum, 7, 15, 16, *16*, 17, 26, *26*, 28

Nagle, Brett, *168*

National Audubon Society, 45. See also Audubon, John James

Native Americans, 15, 38, 65, *220*. See also Dakota peoples; Indigenous peoples; Ojibwe peoples

natural history: art and, 135–47; collections pertaining to, 14, 15, 19, 21, 52, 114, *218*; films on, 106–7; support for, 158. See also Cedar Creek Natural History Area (University of Minnesota); Geological and Natural History Survey Act of 1872 (Minnesota); Minnesota Geological and Natural History Survey

natural history museums: challenges faced by, 186, 196; dioramas and displays, 33–34, 53, 58; early treatment of Indigenous people, 65; education programs, 47–50, 92, 94; mission of, 1, 3, 96; university-based, 114, 186, 187, 189. *See also individual natural history museums*

natural selection, Darwin's theory of, 162, 164

nature, 34, 175; integrating art and science with, 3, 45, 69, 135–47; presenting, 5, 55–56, 94, 109, 114; St. Paul Campus building reflecting, 203–3; studying, 1–3, 74, 82–88, 138, 159, 169, 199–100. See also dioramas; evolution; exhibits; films, nature

Nature Conservancy, 107, 130, 178

Nelson, George "Pinky," 195, *193*

Nerdahl, Rodney, 191, 195

New Guinea Binatang Research Center, 175

Nordquist, Gerda, work with endangered species, 125, 126, *127*

Northrop, Cyrus, 21

Nuechterlein, Gary, 123

Oestlund [Professor], *15*

Ojibwe peoples: language, 222

Oldenburg, Margaret, 156, *156*

Olson, Sigurd, 69

opossums, 166–67, *167*

orioles, *162*

ornithology, 16, 80, 105, 138, 160–63; curators of, 78, 82. See also birds; Breckenridge Chair of Ornithology; Minnesota Ornithologists' Union (MOU)

Otter diorama, *235*

owls, *49*

Ownbey, Gerald, 157

Packard, Jane, 118

Packer, Craig, 199

paddlefish, *17*

paintings. See art, integrating nature and science with; Breckenridge, Walter "Breck," wildlife paintings; Jaques, Francis Lee, wildlife paintings; *and individual artists*

Palisade Cliff, *133*

Palmquist, Dave, 102

parks, state, interpretive programs, 50. See also Itasca State Park

Parmelee, David, polar research by, 74–75, *75*, 115

Parmelee Massif (Antarctica), 75

Passenger Pigeon diorama, 8, *226*, *238*

Pawlenty, Tim, Governor, 187, 188, 189, 192, 194

Peale, Charles Wilson, 33–34

peatlands, exhibit on, 145, *145*

penguins, 75

peregrine falcons, *128*; exhibit on, *9*, *129*, *146*, *147*, 200; return of, 129–33, *130*, *131*, *132*, *133*

pesticides, 129, *130*, 153

Peterson, Roger Tory, 5, 59

Peterson, Sonja, resident artist, *142*

Pettingill, Olin S., Jr, 78

Pfannmuller, Lee, *187*

Philippines, Menage expedition to, *18*, 19

phycology, 21, 23. *See also* algae

phylogenetic trees, 118, *164*, 164–68, *171*. *See also* Tree of Life Project

Pidgeon, Anna, *104*, 104–5, *105*

Pierce, Doug, 204

Pipestone Prairie diorama, 29, 38–39, 42, *43*, 233

planetarium. *See* Minnesota Planetarium

plants: around St. Paul Campus building, 204–3, *204*; collections of, 21, 24, 151, 155–59, 179, *179*, 182; curators of, 169, 172, 175; depicting, 28, 33–34, 58–59; in dioramas, 35, *35*, 38, 39, 62; gene sequencing of, 169–70; interaction with animals and the environment, 34, *220*; mapping, 157–58; rare and endangered, 29, 125–27; succession in, 87. *See also* botany; flora; Herbarium; *and individual plants*

plovers, upland. *See* upland sandpipers

pollination/pollinators, 111, *174*, 175, 204. *See also* honeybee science lab (Honeybee Program)

rabbit, radio-tagging, 83, *83*. *See also* Canada Lynx with Snowshoe Hare diorama

Raccoon diorama, 234

radio telemetry, 2, 74, 80, 82, 82–87, *84*, *85*, 115, *121*

ravens, *118*, 137

Reddy, Sushma, *11*, *170*, 170–71; appointed to Breckenridge Chair of Ornithology, 3, 163

Red Fox diorama, 234

Redig, Patrick, and return of peregrine falcons, 130, *130*, *131*, 133

Regal, Phil, 115, *116*

reptiles: collections of, 19, *116*, 151, 155; curators of, 175; study of, 2, 44, 95, 115, 123, 172. *See also* amphibians; herpetology

research: Antarctic, 74–75, *117*; Arctic, 72–73, *73*, 76–77; Bell Museum's role in, 2–3, 5, 52, 56; in biodiversity, 150, 159, 169, 172–79; field, 45, 74, 102, *117*, 117–18, 121, *121*, 199; funding, 44–45, 83, 84; goal of inclusivity, 3–4; online resources for, 181–82; presenting to public, 101, 108, 135–43, 145–47, 186, *196*, 198–100, 225; University, 56, 99, 186, 189; use of collections for, 150–53. *See also* Bell *LIVE!*; Cedar Creek Natural History Area (University of Minnesota); Delta Waterfowl Research Station (Manitoba, Canada); DNA sequencing; education, graduate studies program; honeybee science lab (Honeybee Program); science; Tree of Life Project; visitors, audience research

Resident Artist Research Project (RARP), 140, *142*

Rich, Andrea, *Bittern*, 138

Richardson, Jenness, 29, 48; dioramas prepared by, *34*, 35–39, 42, 49

Richardson, Olive (wife of Jenness), dioramas prepared by, 35, 38

Ridgway, Robert, 26

Ring-necked Pheasant diorama, 235

Roberts, Jennie Cleveland (wife of Thomas), 26, *27*, 30

Roberts, Thomas Sadler: *The Birds of Minnesota*, 4, 30, 45–46, 50, 69, *69*, 125, *135*, 136–37; bird studies, 4, 25–31, 47, *47*, 55; death of, 31, 44; dioramas prepared by, 34–39, *39*, 53, 55, 135; director of Bell Museum, 4, 15, 27–31, 47, 72; Jaques hired by, 52–53; medical practice, 26, 27–28, 55; nature films produced by, 28–29, 43, 48–49, 106, 109; photographs of, *4*, *7*, *26*, *27*, *28*, *29*, *30*, *31*, *48*; public programs launched by, 47–49, 51; Walter Breckenridge hired by, 42; work for new natural history museum building, 30–31, 55–56, *57*

Roe, Areca, resident artist, *142*

Rosendahl, Carl Otto (C. O.), 23–24, 155, *157*

Rubens, H. W., Heron Lake diorama paintings, 38

Rudnick, Larry, 192, *190*

Rysgaard, George, 98

Sagan, Carl, 142

salamanders, 175–76, *176*, *179*

Sandberg, John, 155

sandhill cranes, *43*, 43–44

Sandhill Cranes diorama, 31, 61, *61*, 101, *222*, *234*; moving, *211*

sandpipers. *See* upland sandpipers

Sargeant, Alan, 85, *85*

Saturday with a Scientist program, 97, *97*

schools. *See* dioramas, circulating to schools; education, school groups

Schott Learning Kit program, 95

Schultz, Vincent, 83

science: Bell collections' value to, 152–53; education in, 110–12; integrating art and nature with, 3, 45, 69, 135–47; natural, 186; women in, 21–24, 156. *See also* citizen scientists; *and specific fields of science*

seabirds, 73, 75

seal, 74; radio-tagging, 85

Seal, Ulysses, 115

Self, Ruth, *62*

Shuman, Bryan, 108, 199

Sibley, David, 5

Simons, Andrew, *152*, 167–68, *168*, 169

Siniff, Don, 74, 84, *85*

Sloan, Bob, 115

Smallmouth and Largemouth Bass diorama, *236*

Smith, Welby, 157–58, *158*, 159

Smithsonian Institution (Washington, DC), 14, 92, 144

snakes, venom of, 166–67

Snow Geese diorama, 31, *60*, 61, *232*; moving, 211, *212*

sparrows, *47*, *135*, 161

species, 151, 160, 171; conservation of, 3, 30, 44, 45; formation of, 164, 176; identifying, 16, 26, 152, 158; invasive, 145, 146; loss of, 172, 175; Minnesota native, 38, 64; relatedness of, 99, 115, *164*, 165; studying, 74, 117, 120–21. *See also* animals, rare and endangered; biodiversity; Endangered Species Acts; plants, rare and endangered

Spilhaus, Athelstan, 82

Spivak, Maria, 111–12, *112*

Stanton, Daniel, 171, *171*

Star Man (puppet), 142

succession, 87, 171

Sustainable Shelter (exhibit), *147*

Sutton, George Miksch, 74, 75, 78, *78*, 136

Swallow-tailed Kite diorama, *235*

swans, 10, 138. *See also* Tundra Swans diorama

systematics, 150, 160, 163

taxidermy, 34, 35, 42, 53, 58, 62

Taylor, Philip, on Arctic research expedition, 72

Tester, John, 45, 83, 121; Canadian toad project, 86, 86–87

Thomas, Donald, resident artist, *142*

Thompson, Rolf, 102

Tilden, Josephine, 15, *21*, 21–24, 155

Timm, Bob, 116

Tinbergen, Nikolaas, 120

toads, Canadian, 86, 86–87

Tordoff, Harrison "Bud," 114–18, *115*; director of Bell Minnesota, 3, *114*, 186; and return of peregrine falcons, 9, 125, *129*, 129–33, *130*, *131*

Touch and See Room/Lab, 9, 51, 90, 90–93, *91*, *92*, *93*, 99, *104*, 213. *See also* Barthelemy, Richard "Bart"

Tree of Life Project, *164*, 164–68, 169–71, *218*. *See also* phylogenetic trees

Troubled Waters: A Mississippi River Story (documentary), 108

Tundra Swans diorama, 62, *233*; moving, 213

Underhill, Jim, work with endangered species, 125, 126

University of Minnesota (U of M): Animal Biology Building, 28, 30; Bell Museum's connection to, 2–3, 145–46, 197, *196*, 198–100; Biology Colloquium, 75; center for natural history study in Minnesota, 14, 26–27; mission, 2, 159, 186; museum studies graduate minor, 95; Old Main building, 1, 2, *2*, 6, 15, *15*, 155; Pillsbury Hall, 6; radio station, 49; research at, 56, 99, 186, 189; Zoological Museum, 15, 28, 30, 47; Zoology Building, *7*, 55, 56. *See also* Cedar Creek Natural History Area (University of Minnesota);

Lake Itasca Forestry and Biological Station (University of Minnesota); Minnesota Museum of Natural History; Minnesota Seaside Station (Vancouver Island, Canada)

University of Minnesota Herbarium. *See* Herbarium

upland sandpipers, 125; Wong's drawing of, *9, 124*

vangas (birds), 170–71, *171*

Vazquez, Hernan, *168*

Villa Ramírez, Bernardo, 78–79

visitors: audience research, 51, 196–97, 198; engaging, 90, 94–98, 142, 144, 147, 197–98, 222, 227; interacting with exhibits, 51, *51, 146, 224, 227*; to St. Paul Campus building, 198–100, *198*; virtual, 182

Wachter, Joseph, 24

walruses, 73

Warner, Dwain Willard, 45, 75, 78, 78–81, *79, 80*; radio telemetry work, 2, 80, *82,* 82–85

Washburn-Crosby Milling Company, 53. *See also* Bell, James Stroud; General Mills Corporation

Waubun Prairie, 86–87, *87*

Weaver Dunes, *131*

Weber, Walter, *The Birds of Minnesota* plates paintings, *135,* 136

Webster, William and Byron, 138

Weiblen, George, 4, *169,* 172, *173, 174,* 175, 205

Weller, Susan, director Bell Museum, 11, *11, 139,* 163, 186, 188, *186*

West, Payton, 199

Wetmore, Clifford, 181

White-tailed Deer diorama, 34, *230*

Whitfeld, Tim, *159*

Wilcox, Arthur, 82

wildlife, 39, 58; Arctic and Antarctic, 73, 74; art of, 45, 137, 147; depicting, 33, 34, 62, 64–65, 69; filming, 106, *107*; populations of, 26, 30; radio-tagging, 74, 80, 82–87; refuges for, 3, 38; study of, 2, 76–77, 101. *See also* animals; fauna; Minnesota Geological and Natural History Survey; *and individual wildlife species*

Wilkie, Robert and James, on Arctic research expedition, 72, *72,* 73

Williams, Kevin, *96,* 110, *110,* 146

Wilson, Alexander, wildlife paintings by, 138

Wilson, E. O., *Sociobiology: The New Synthesis,* 121

Wilson, Michael, resident artist, 140

Winchell, Horace, and Menage expedition, 19

Winchell, Newton Horace, 1–2, 5, *6, 14*; Geological Survey and Zoological Museum director, 14–16, 26

Windows to Nature: Minnesota's Dioramas (documentary), 108

Winston, William O., 135

Wolves diorama, *7, 8,* 31, 55, *56,* 58–59, *59, 98, 222, 233*; moving, 211

women, in science, 21–24, 156. *See also individual women*

Wong, Vera Ming, *Upland Sandpiper, 9, 124*

Worcester, Dean, and Menage expedition, *18,* 19

Wunderle, Joe, 117

Yang, Ya, *169,* 169–70, 171 Young, Denise, director Bell Museum, 11, 229

Young, Denise, director Bell Museum, 11, 229

Young Naturalists' Society, 26

Zaca expedition, 68

Zink, Robert: appointed to the Breckenridge Chair of Ornithology, 160–63, *161,* 169

Zoological Museum, 6, 7. *See also* Bell Museum; General Museum; James Ford Bell Museum of Natural History; Minnesota Museum of Natural History; Museum of Natural History

Zoological Survey. *See* Minnesota Geological and Natural History Survey

zoology: collections pertaining to, 15, *15, 24*; courses in and faculty for, 2, 14, 16, 23, 34, 78, 114–15, 121; curator of, 28

The Bell Museum

The Bell Museum, Minnesota's official natural history museum, was established by the state legislature at the University of Minnesota in 1872. For more than a century, the museum has preserved, studied, and interpreted the state's rich natural history and served learners of all ages. Its scientific collections contain more than one million specimens representing every county in Minnesota and thousands of locations worldwide. Its art and film collections include the rare *Birds of America* double elephant folio by John James Audubon and the classic documentary *Minnesota: A History of the Land*.

The mission of the Bell Museum is to ignite curiosity and wonder, explore connections to nature and the universe, and create a better future for our evolving world. Learn more about the museum, including information about events and programs, at bellmuseum.umn.edu.

Minnesota Journeys

The permanent galleries, which include world-renowned wildlife dioramas, span space and time from the origins of the universe, through the diversification of life on Earth, to Minnesota's unique habitats. The exhibits connect visitors to scientists, nature, and the search for a sustainable future.

Whitney and Elizabeth MacMillan Planetarium

The state-of-the-art planetarium takes visitors on amazing journeys from the far reaches of the cosmos to deep inside the human brain. The ExploraDome, a traveling planetarium, reaches schools and community centers across the state.

Touch & See Lab

The Bell Museum created the first natural history discovery room in the world, a tradition continued in the Touch & See Lab, where all ages actively learn through observation and sensory engagement.

Learning Landscape

Learning continues outside the museum with sustainable landscaping, water features, native plants, art installations, geology exhibits, and an observation deck.

Mnisóta Makhóche

The Bell Museum is located on the traditional and treaty land of the Dakota people, in the area known as Mnisóta Makhóche (the land of Minnesota).